Construction Process Improvement

Construction Process Improvement

Edited by

Brian Atkin, Jan Borgbrant & Per-Erik Josephson

Competitive Building

Blackwell
Science

Editorial Offices:
9600 Garsington Road, Oxford OX4 2DQ, UK
Tel: +44 (0)1865 776868
Blackwell Science, Inc., 350 Main Street, Malden, MA 02148-5018, USA
Tel: +1 781 388 8250
Iowa State Press, a Blackwell Publishing Company, 2121 State Avenue, Ames, Iowa 50014-8300, USA
Tel: +1 515 292 0140
Blackwell Publishing Asia Pty Ltd, 550 Swanston Street, Carlton South, Victoria 3053, Australia
Tel: +61 (0)3 9347 0300
Blackwell Wissenschafts Verlag, Kurfürstendamm 57, 10707 Berlin, Germany
Tel: +49 (0)30 32 79 060

First published 2003 by Blackwell Science Ltd

Library of Congress
Cataloging-in-Publication Data
Construction process improvement/edited by
Brian Atkin, Jan Borgbrant & Per-Erik Josephson.
p. cm.
Includes bibliographical references and index.
ISBN 0-632-06462-5 (softcover)
1. Building–Cost effectiveness. I. Atkin, Brian.
II. Borgbrant, Jan. III. Josephson, Per-Erik.

TH438.15.C65 2003
690′.068–dc21 2002038566

ISBN 0-632-06462-5

A catalogue record for this title is available from the British Library

Set in 10/13 pt Palatino
by Sparks Computer Solutions Ltd, Oxford
http://www.sparks.co.uk
Printed and bound in Great Britain by
MPG Books Ltd, Bodmin, Cornwall

For further information on Blackwell Science, visit our website:
www.blackwell-science.com

Contents

Preface xi
Acknowledgements xiii
Contributors xv

1 Introduction **1**
Brian Atkin, Jan Borgbrant & Per-Erik Josephson
Recognising the importance of process 1
Creating an effective response 1
Balancing concern for the product with the process 2
The process to be improved 3
Aims of this book 5
Topics addressed 6
Scope and content 7

2 Modularisation in the Customisation of Manufactured Housing **15**
Carina Johansson
Introduction 15
State-of-the-art review 16
Research project 25
Research results and industrial impact 27
Conclusions 28
References 29

3 Application of Integrated Life Cycle Design to Housing **31**
Mats Öberg
Introduction 31
State-of-the-art review 32
Research project 37
Case study: research results and impact 38
Conclusions 41
References 41

4 **Life Cycle Costs of Commercial Buildings – a Case Study** 44
Eva Sterner
Introduction 44
Methodological considerations 45
Case study description 46
Energy conservation 48
Life cycle cost analyses 50
Conclusions 54
References 55

5 **A Life Cycle Cost Approach to Optimising Indoor Climate Systems** 56
Dennis Johansson & Anders Svensson
Introduction 56
State-of-the-art review 56
Research project 62
Research results and industrial impact 64
Conclusions 65
References 66

6 **Performance Indicators as a Tool for Decisions in the Construction Process** 68
Veronica Yverås
Introduction 68
State-of-the-art review 68
Research project 74
Research results and industrial impact 77
Conclusions 79
References 80

7 **Reducing the Risk of Failure in Performance within Buildings** 82
Stephen Burke
Introduction 82
State-of-the-art review 82
Research project 89
Research results and industrial impact 90
Conclusions 91
References 91

8 **Physical Status of Existing Buildings and their Components with the Emphasis on Future Emissions** 93
Torbjörn Hall
Introduction 93

State-of-the-art review 94
Research project 101
Research results and industrial impact 103
Conclusions 104
References 104

9 Co-ordination of the Design and Building Process for Optimal Building Performance 106
Niklas Sörensen
Introduction 106
State-of-the-art review 107
Research project 111
Conclusions 115
References 116

10 New Concrete Materials Technology for Competitive Construction 118
Markus Peterson
Introduction 118
State-of-the-art review 118
Research project 124
Research results and industrial impact 125
Conclusions 127
References 127

11 Competitiveness in the Context of Procurement 130
Fredrik Malmberg
Introduction 130
State-of-the-art review 131
Research project 140
Conclusions 141
References 141

12 Encouraging Innovation through New Approaches to Procurement 143
Kristian Widén
Introduction 143
State-of-the-art review 143
Research project 149
Research results and industrial impact 150
Conclusions 151
References 151

13 **Public-Private Partnerships – Conditions for Innovation and Project Success** **154**
Roine Leiringer
Introduction 154
State-of-the-art review 155
Research project 160
Research results and industrial impact 163
Conclusions 165
References 166

14 **Pros and Cons in Partnering Structures** **168**
Anna Rhodin
Introduction 168
State-of-the-art review 168
Research project 176
Research results and industrial impact 178
Conclusions 180
References 180

15 **Importance of the Project Team to the Creation of Learning Within and Between Construction Projects** **183**
Fredrik Anheim
Introduction 183
State-of-the-art review 184
Research project 188
Research results and industrial impact 190
Conclusions 192
References 193

16 **Refurbishment of Commercial Buildings: the Relationship between the Project and its Context** **195**
Åsa Engwall
Introduction 195
Theoretical framework 196
The case study of Oxenstiernan 201
Discussion and conclusions 208
References 209

17 **Improving Project Efficiency through Process Transparency in Management Information Systems** **211**
Christian Lindfors
Introduction 211
State-of-the-art review 212
Research project 221

Conclusions 222
References 223

18 Improvement Processes in Construction Companies 225
Peter Samuelsson
Introduction 225
State-of-the-art review 226
Research project 230
Research results and industrial impact 232
Conclusions 237
References 238

19 Design Research and the Records of Architectural Design: Expanding the Foundations of Design Tool Development 240
Robert Fekete
Introduction 240
State-of-the-art review 240
Conclusions 253
References 253

20 Communicating Project Concepts and Creating Decision Support from CAAD 255
Jan Henrichsén
Introduction 255
State-of-the-art review 255
Research project 260
Research results and industrial impact 264
Conclusions 264
References 265

21 Using 4D CAD in the Design and Management of Vertical Extensions to Existing Buildings 266
Susan Bergsten
Introduction 266
State-of-the-art review 267
Research project 274
Research results and industrial impact 275
Conclusions 275
References 276

22 Importance of Architectural Attributes in Facilities Management 278
Ulf Nordwall
Introduction 278

State-of-the-art review 279
Research project 285
Research results 287
Conclusions 290
References 290

23 Conclusions **292**
Brian Atkin, Jan Borgbrant & Per-Erik Josephson
Ways forward to construction process improvement 292

Index 299

Preface

Over the past few years, researchers and sponsors have increasingly turned their attention to finding ways of improving the construction process. After decades of neglect, construction process is high on the agenda and will remain so, because of the realisation that the scope for improvement is considerable and the time to make a difference in practice is long. Indeed, there are no quick fixes. Sustained improvement can come about only through well-conceived plans that find acceptance among practitioners, not least clients. Moreover, concern for the process has to be matched by concern for the product. Considering one without the other is unlikely to lead to breakthroughs.

In this connection, new management thinking and actions that ignore the underlying technology and how it is designed, produced, delivered, incorporated in a building and then maintained will have limited impact. Inevitably, some new thinking will prove to be lacking, with the risk of a legacy of failed buildings. For these reasons, the subjects covered by this book are an attempt to ensure that product-related questions are not overlooked in the search for answers to improvement in the construction process.

Twenty-one chapters cover many different aspects of the construction process and, in most cases, explicitly cover their product-related implications. Even so, they represent a fraction of the subjects that could or might have been included. The problems to be solved are too numerous to mention, but, at least, some important steps have been taken and are described by the contributors. The key directions given to the contributors were: 1) to set their work in its proper context, and 2) to ensure that the methodological aspects were defensible and transparent. *Construction Process Improvement* represents many strands in the search for better value and a more enduring built environment and, as such, should appeal to readers with a more holistic concern for the process.

Brian Atkin, Jan Borgbrant & Per-Erik Josephson
Reading, Luleå & Gothenburg

Acknowledgements

Competitive Building is a five-year, 70 million SEK (€7.6 million) collaborative research and development programme. The Swedish Foundation for Strategic Research (SSF) has provided a significant proportion of this total, with the balance contributed by the Development Fund of the Swedish Construction Industry (SBUF); the Swedish Council for Environment, Agricultural Sciences and Spatial Planning (FORMAS, incorporating BFR); and the Swedish construction sector, in the form of individual companies. We thank them for their support and, with it, the opportunity to prepare this book.

Contributors

The Editors

Brian Atkin PhD, MPhil, BSc, FRICS, FCIOB is Programme Director for the Swedish national construction R&D programme, *Competitive Building*, a position he holds through Lund Institute of Technology, part of Lund University. He holds or has held professorial appointments at the University of Reading in the UK, VTT (Technical Research Centre of Finland), Helsinki University of Technology, Royal Institute of Technology in Stockholm, Chalmers University of Technology in Gothenburg and the University of Hong Kong. He is a Director of Atkin Research & Development Limited, a specialist consultancy, and AECademy Limited, an internet distance learning company. He is joint author of *Total Facilities Management*, also published by Blackwell.

Jan Borgbrant PhD is Professor and Vice President of the University of Luleå and has held a professorship at Lund Institute of Technology. He is Vice Programme Director for *Competitive Building*. From 1980, he was senior researcher at the Swedish Council for Management and Organizational Behaviour, head of the administrative division at the National Board of Social Security, senior consultant at the Swedish Institute of Management and Senior Vice President of the construction company BPA. He is presently Chairman of the Board of Sikta Management AB and the Swedish Institute of Building Documentation.

Per-Erik Josephson PhD is a Docent and Associate Professor at the Department of Building Economics and Management, Chalmers University of Technology in Gothenburg, and a member of the management group for *Competitive Building*. His research covers project management, quality management and change management. He developed and initiated the masters degree programme in civil engineering and architectural management at Chalmers, for which he received an award for excellence in teaching. He is the Principal of a project management consultancy business.

Contributors

Fredrik Anheim, Luleå University of Technology and NCC AB
Susan Bergsten, Luleå University of Technology and Swedish Institute of Steel Construction
Stephen Burke, Lund Institute of Technology
Åsa Engwall, Luleå University of Technology
Robert Fekete, Lund Institute of Technology
Torbjörn Hall, Chalmers University of Technology and White Arkitekter AB
Jan Henrichsén, Lund Institute of Technology
Carina Johansson, The Royal Institute of Technology and Peab AB
Dennis Johansson & **Anders Svensson**, Lund Institute of Technology and Stifab Farex
Roine Leiringer, The Royal Institute of Technology
Christian Lindfors, The Royal Institute of Technology and NCC AB
Fredrik Malmberg, Lund Institute of Technology and Malmö-Lund Master Builders Association
Ulf Nordwall, Chalmers University of Technology and Bostads AB Poseidon
Mats Öberg, Lund Institute of Technology and Cementa AB
Markus Peterson, Lund Institute of Technology and Skanska AB
Anna Rhodin, Luleå University of Technology and Skanska AB
Peter Samuelsson, Chalmers University of Technology and Skanska AB
Eva Sterner, Luleå University of Technology
Niklas Sörensen, The Royal Institute of Technology and Scandiaconsult
Kristian Widén, Lund Institute of Technology and Malmö-Lund Master Builders Association
Veronica Yverås, Chalmers University of Technology and JM AB

Chapter 1
Introduction

Brian Atkin, Jan Borgbrant & Per-Erik Josephson

Recognising the importance of process

Modernising the construction sector is a declared aim of many governments and agencies. Long seen as inefficient and uncompetitive, the construction process – taken in its broadest sense – has become the target for research projects and programmes in many countries. One such programme is Sweden's national R&D programme in construction, *Competitive Building*.

Most research activity has been aimed at the products of construction or the science and technology that helps create and maintain them. Despite the considerable investments in building research, many problems remain, costs continue to rise and customers – building owners and occupants – complain of a sector that serves them badly. This is a familiar theme and one that has led to a partnership between universities and industry aimed at improving the efficiency of the construction process.

The mandate for *Competitive Building* was clear: improve the process and raise construction's competitiveness. The implementation was less straightforward. Construction process is one branch of building research that has been neglected for decades. Placing the case for process-related research high on the agenda was a task that fell to the leading companies and firms in the sector, supported by a group of senior researchers from the main science and technology universities. That case was made and accepted in 1997 against a detailed programme proposal for collaborative R&D that would address the construction process. The programme began in earnest the following year.

Creating an effective response

The strategic aims of *Competitive Building* are to help develop the competence of the sector's workforce and to improve industrial competitiveness through a more efficient process. It achieves these aims through the active co-ordination of research projects and the provision of postgraduate, researcher education. Specific objectives are to:

(1) Develop and renew researcher education within the building sector.
(2) Act as a driver for the sector's strategically important research projects.
(3) Develop an effective network between the sector's companies and the universities.

These aims and objectives reflect the many challenges facing the sector, including the expectations of society, as well as the concerns of those engaged within the sector:

- creating buildings with a *healthy indoor climate*;
- meeting the goal of *sustainable development* through a life-cycle approach;
- improving *architectural quality* in order to create excellent buildings;
- increasing *productivity and industrialisation*; and
- adopting a *market-controlled and customer-oriented* process.

Success in future markets is foreseen to come from superior products that can be delivered to demanding environmental, quality, cost and time targets. In order to achieve these objectives, there has to be a process to fit. *Competitive Building* has been set the task, amongst other things, of providing the focus and stimulation for research that will have direct industrial application. Furthermore, it intends to avoid the risk of missed opportunities for improvement by placing strategically important research projects within companies. In practice, this means combining theory with practice in the pursuit of research project objectives and developing *agents of change*, i.e. industry-aware researchers who are employed by companies to help in implementing the results of their collaborative research efforts.

Thirty researchers have been recruited to join *Competitive Building*'s competence-developing programme to focus on resolving a problem of significance to an individual company, as well as the sector at large. They are supported in their work by structured, progressive research education.

Balancing concern for the product with the process

The fields included in *Competitive Building* are many, reflecting several branches of building research. One thing has become clear and that is a focus on process alone will fail to deliver the advocated changes. Process and product are inextricably linked, so that consideration of one in isolation from the other is a recipe for failure. There have been many attempts at bringing about change in the way buildings are produced, only to fall by the wayside because they ignored the product and what it meant to those who had to use those buildings. Housing is a prime case. Another example is that of novel products and systems that are designed and fabricated away from any concern for how they might be shipped, handled, installed and maintained. One could fill a book on examples of failures arising from a lack of integration of product and process.

Competitive Building is an attempt to address this imbalance by targeting process improvement, but not in isolation from the product. Many of the projects within the programme have an obvious attachment to some aspect of the sector's products, and as will be shown in subsequent chapters, there are strong arguments for handling process and product in this way. One of the concerns recognised by *Competitive Building*'s programme management is the risk of working with research results that are too generic or removed from the needs of industry. A distinct feature of the programme is, therefore, that the projects, as described and discussed in this book, are a mixture of technical- and human-oriented topics and of strategic and operational targets. A national R&D programme has to have clear, strategic goals or targets, but it also has to base its workplan on a portfolio of projects that can cover sufficient problems to achieve results that will impact upon the sector, as well as help to spread risk. At the practical level, the projects interact through scientific debate, enquiry and the sharing of experience and knowledge. Much of the latter is due to the efforts of individual researchers – the contributors to this book – and their co-workers.

The process to be improved

Concern for the process and how it is expected to align with the demands increasingly placed upon it has led to its redefinition. This is shown in Fig. 1.1 in a form that will be familiar to anyone engaged in process improvement. It emphasises three product-related phases: definition, manufacture and use. There is, therefore, no claim as to originality in this thinking. However, it does predate many initiatives that have launched into action in recent times. Its real importance, even its strength, is that those who recognise the problems afflicting the sector, and who are in a position to do something about them, accept it. In this way, the figure has provided a useful starting point, as well as a common point of reference, for the workplan of *Competitive Building*.

Competitive Building's projects are grouped under two themes: industrialised housing for good living and rationalised real estate redevelopment. Within them are thirty projects of which two thirds are featured here.

Industrialised building for good living (Fig. 1.2) covers:

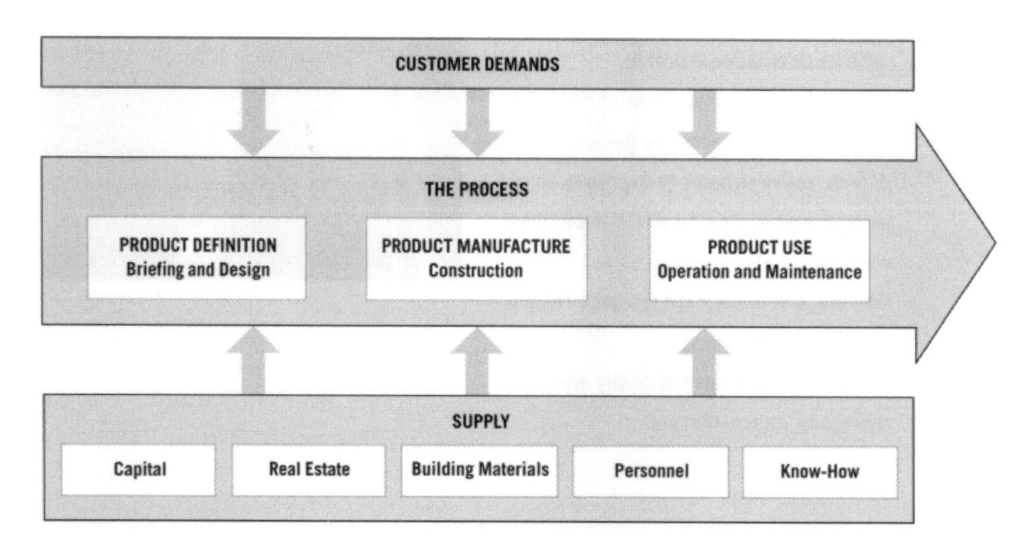

Fig. 1.1 A process view of improvement.

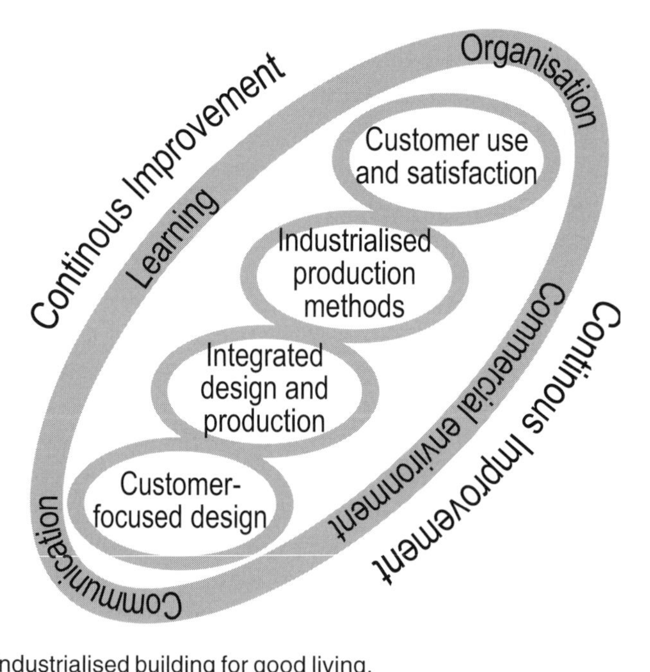

Fig. 1.2 Industrialised building for good living.

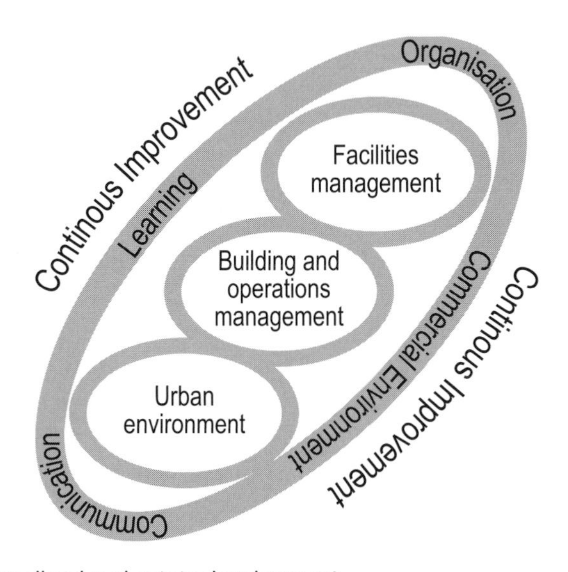

Fig. 1.3 Rationalised real estate development.

- customer-focused design;
- integrated design and production;
- industrialised production methods; and
- customer use and satisfaction.

Rationalised real estate redevelopment (Fig. 1.3) covers:

- facilities management;
- building and operations management; and
- commercial environment, communication and learning.

Aims of this book

An aim of this book is to disseminate the results of current research within *Competitive Building* and, therefore, bring it to the attention of a wide readership. The intention has been to create a body of knowledge that provides a solid point of reference for researchers and enquiring practitioners who wish to delve into areas that might have application to their work. Each chapter is devoted to a topic that addresses a problem defined by the researcher and his or her collaborating company. This is supported in most cases – depending upon the stage the research has reached – by a description of the research project, including methodological treatment, tentative results and assessment of industrial impact. Each chapter contains a literature review, which in many cases is contained within a separate section. From this review, it is possible to see where problems exist, where there are gaps in our knowledge and where there is a good case to be made for the research in question.

None of these individual contributions claims to solve major problems, though collectively they represent a serious and targeted research action that is relevant to the needs of industry. What they do provide is a transparent account of the approach the researchers are taking to their work. They have been very open with these accounts, seeing the need to show why they are going about their research in the way they are. One of the concerns that many researchers and practitioners share is the lack of reproducibility of research results. In the natural and physical sciences, it is necessary to show clearly how the results were arrived at; in the medical sciences, concealment can have grave consequences. Though we would hesitate to place building research on quite the same scientific level, we must, nonetheless, aspire to it. Buildings can bring their own serious consequences, if the product and the process that delivers it are left unchecked.

Reproducibility of research results is one of ten criteria that we use to define *good* research:

(1) necessity and sufficiency;
(2) contextual awareness;
(3) knowledge of the literature;
(4) expression of the problem and goal(s);
(5) methodological transparency;
(6) rigorous application of method(s);
(7) data integrity;
(8) reproducibility of the results;
(9) clarity of communication; and
(10) peer-group acceptance.

These are the ideals to which any researcher should subscribe. They have, therefore, been used to frame and assess the contributions in this book.

Topics addressed

The topics that form the basis for each of the subsequent chapters are drawn directly from the projects within *Competitive Building* with the inclusion of one other that has strong associations with the programme's aims and workplan. Each research project addresses one or more questions of strategic importance, even where the results are expected to contribute to operational improvement. The context of operational problems cannot be considered in isolation. It must relate to some strategic purpose or objective for a company or, better still, the sector at large.

All the research projects described and discussed in subsequent chapters are the subject of university–industry collaboration. In most cases, industry is represented by a single enterprise and in others it is a consortium. Many are construction companies, but others are design firms, material producers and, importantly, building owners. Reading of the individual accounts will provide insights into the very real problems that successful companies face. They are not pursuing research goals because they are doing badly. They are aiming to improve their efficiency and competitiveness. Since they constitute some of the major players in the sector, their influence can be felt widely at home and also, in some cases, internationally.

Scope and content

The chapters that follow are arranged in a sequence that adheres largely to the phases in the construction process. There has been no intention to shoehorn any of the contributions into a prescribed slot. Each contribution is there because it has something valuable to disseminate and because it taps sources of literature that draw from many fields, experiences, cultures and, even, languages. This book, therefore, makes a distinct contribution by having accessed literature that other researchers and practitioners will hopefully find useful.

In our first main chapter, *Carina Johansson* addresses the production of customised products from standardised parts, which is a common approach in many sectors in order to create variants of products that satisfy different customers' needs. To become competitive and remain so, especially in international markets, the house-building sector faces the challenge of, amongst other things, creating design concepts that take advantage of advanced manufacturing methods and that are adaptable to local conditions. Adaptability to local conditions requires knowledge of building traditions and regulations, owners' and occupiers' preferences and the whole life costing of the building for the countries affected. A modularisation approach is being adopted to identify how product variance, components' life cycles, maintenance and replacement costs and intervals can be used to create concepts for modular building systems that are adaptable to different countries.

Mats Öberg investigates design and procurement based on whole life appraisal. This is seen as an important step towards improving functional quality and, therefore, the overall cost-effectiveness of buildings. Empirical studies support this claim, yet only a small proportion of building projects actually adopt life-cycle design principles. In this regard, whole life appraisal is examined by considering design procedures and the nature of the product. The primary approach adopts life-cycle costing, life-cycle assessment and service-life planning within the concept of integrated life-cycle design. *Open building* may be regarded as the resulting product and a review of the wider benefits are considered. A study of multi-family dwellings has explored the practicability of this approach in terms of the application of life-cycle thinking during design, life-cycle cost estimating and accuracy in predicting energy use.

Eva Sterner shows that for construction projects with high environmental targets, life-cycle cost analyses are useful tools for comparing designs and for justifying construction and operational strategies from an economic perspective. Here, a case study of three buildings reveals how environmental strategies can affect life-cycle costs. It also shows the importance of a long-term economic perspective on investment appraisal. Initial, maintenance and operational costs are established and then compared with the life-cycle costs of conventional buildings in which environmental attributes are absent. The results show that the conventional buildings attract the lowest initial cost, but that operational energy use is significantly higher than in the environmentally-designed buildings.

Dennis Johansson & Anders Svensson discuss the need for the indoor climate of a building to be appropriate to the intended function of the building. They stress that it should be of the lowest total cost over the whole life-cycle. Sometimes it is not. The use of life-cycle cost (LCC) analysis on new and refurbished buildings is low, especially regarding indoor climate systems. Reasons for this situation include the absence of technical know-how and tools with which to perform LCC analyses. Research is addressing the optimisation of indoor climate systems from a whole life perspective since an energy efficient system decreases total cost of ownership over the life cycle. Energy efficient systems are also more flexible and stable over time, offering better functionality and a healthier workplace.

Veronica Yverås tackles the concern that many people share for the quality of the built environment. The media are also sensitive to this concern and are fond of headlining problems affecting buildings. Part of the justification is a belief that problems are avoidable, perhaps because the same or similar problems are seen to recur. Knowledge on how to avoid problems is available, but is not being applied consistently. Methods with which to compare different design solutions and foresee the incompatibility of materials, components and systems are lacking. Tools that designers and others can use to assess the performance of a building and its components are therefore needed. Requirements such as a healthy indoor environment, comfort and energy efficiency should go without saying. However, estimating various characteristics to provide easy-to-use and reliable performance indicators during design is missing. The research is being directed towards developing appropriate tools.

Stephen Burke argues that architects, engineers and clients often have conflicting ideas of what should be considered in the design of a new building. These ideas invariably lead to some compromise between the demands of hard engineering and softer issues, with the potential likewise to compromise on the physical characteristics and performance of the building leading to some measure of failure. Examples of failure include high energy costs, health problems and structural damage due to moisture, for which the occupant must pay directly or indirectly. These can also have long-term socio-economic consequences. Current problems and failures resulting from the neglect of building physics principles are examined and their causes are highlighted. Research is continuing into the development of tools to help reduce the risks of failure and to highlight the costs and risks attached to the neglect of building physics principles.

Torbjörn Hall reminds us that a significant proportion of future building activities in many countries will cover the maintenance, upgrading and rebuilding of the existing building stock. This requires knowledge of the conditions inside buildings and tools for assessment of their state and that of components, as well as for supporting subsequent design and management activity. These assessments and tools will, to a large extent, be different from those used for the design and construction of new buildings. However, some knowledge will be common to both kinds of building. Before a decision can be taken on whether to retain, repair or replace components in an existing building, the physical status of those components must be evaluated. The term component covers building materials, structural components and services installations and, therefore, includes entities with generically different characteristics. Research is examining methods of evaluation that take account of environmental conditions, especially those concerning potential emissions of various types.

Niklas Sörensen examines the problems that arise because the construction process is fragmented into different technical disciplines across different phases. This leads to a process in which almost every decision is a compromise between two or more actors over various aspects of the process and the product. Such compromises place heavy demands on co-ordination due to the complex combination of different technology and systems, each of which may have to interact in a different way within the building. The primary focus of attention in the design phase of the process should be the operational phase of the building and is something that should continue throughout the subsequent construction phase. In order to integrate the work of the technical disciplines across the different phases, key technical criteria must be established. These help to pinpoint errors and their cause – for example, a structural problem stemming from an incorrect mix of competence or an economic problem arising from an imbalance between short-term investment cost and long-term operational cost. A study is investigating the causes of potential technical errors and communication breakdowns and some results are presented.

Markus Peterson addresses concerns over the use of cast in situ concrete, which is widely used in the construction of structural frames for multi-storey housing. Traditional low-grade concrete, as typically used in housing, limits floor spans and, in combination with concrete partition walls, reduces flexibility and refurbishment potential. There are also the added complications of moisture control, with long drying periods affecting production time and cost.

Conservatism in the organisation and management of housing contracts also means that novel approaches to the use of concrete have been limited. Over the past decade, advances in materials technology, for instance self-compacting and high performance concrete, have taken place. However, much of the related research concentrates upon the technical properties of concrete, with few scientific results covering design, production and economic aspects. Research is addressing the potential for self-compacting and high-performance concrete and the design and production techniques that are required to create economical solutions that compete well with other materials. Technical and process-oriented aspects are analysed.

Fredrik Malmberg discusses the concept of competition from a client or customer perspective, in terms of the search for best value. In recent years, the number of different procurement methods has increased markedly in response to this search. There is, therefore, no longer a clear or obvious link between the lowest price among tenders and the one that might be described as the best. Newer procurement methods have tended to adopt a more holistic view to the assessment of tenders, bringing in qualitative methods to identify the best offer. An essential element in such assessments is defining the means for measuring best value. This requires that *value to the client* can be properly expressed. Research is examining the extent to which different procurement methods satisfy the call for best value in terms of how well they enable clients to determine the winning tender. Use is being made of a selection of case study projects, both as sources of data and, subsequently, for validation of the research findings.

Kristian Widén contends that the lack of communication and learning inherent in traditional approaches to building procurement and co-operation impacts on the extent of innovation that is possible. New methods and technology do not always find their way into the next project that could benefit from them. The low involvement of suppliers and subcontractors is held partly to blame and is a problem that is unlikely to be overcome until radical changes are introduced to the procurement process. Other changes, such as instilling a culture of innovation within organisations involved in the process, are also needed. A realignment of roles and objectives is required and research is being directed to these aspects, as well as the wider contractual framework. Research is examining the factors that lead to successful innovation against present procurement and contractual arrangements. A new framework for procurement that benefits from a continuous pursuit of innovation is presented.

Roine Leiringer highlights that funding for projects that cater for social needs is becoming increasingly difficult for governments to find. Projects have been put on hold, as traditional public procurement strategies no longer suffice. The search for alternative procurement methods has led to the emergence of public-private partnerships. For the construction sector, these kinds of projects can involve new ways of conducting work, with shifts in responsibilities and risks to the private sector. Much has been claimed as to the benefits of PPP procurement, including lower project costs, shorter construction times and higher overall quality in the end product. Research is examining the design and construction phases for evidence of enablers and inhibitors of innovation, against the goal of project success measured in terms of an improvement over traditional procurement. Findings stress the need for clear and practicable guidelines for construction companies and specialist firms entering into PPP arrangements.

Anna Rhodin finds that the literature on partnering covers many examples of empirical work relating to generally large and complex projects mostly in the US, UK and Australia. Partnering in other settings receives less treatment, but is considered to be no less valid. Research is examining the grounds for adopting partnering arrangements on smaller, more conventional project types as found in many countries. Partnering literature is critically examined and used to pinpoint the arguments that, on the face of it, exclude smaller projects and, conversely, those arguments that might support its adoption. The findings are supplemented by the results of empirical research based on three projects and used to establish a more complete assessment of partnering. The focus is on how the selection process, contract conditions and external environment encourage or inhibit the development of effective collaborative relationships in small- to medium-scale multi-partner arrangements initiated by the client.

Fredrik Anheim argues that, compared with other industrial sectors, the construction sector has experienced consistently rising costs over many years. During the 1990s, a number of actions were taken to reduce the high cost level, including a government-level enquiry. Continuous learning could be one solution in efforts to increase the profitability of companies in the construction sector. This chapter examines the importance of the project team with respect to learning within and between projects. The results of case studies show that building contractors are poor learners. They do not take advantage of the potential learning effects that are provided through teamwork; furthermore, they do not set common goals, nor do they apply dialogue or reflection.

Åsa Engwall finds that in the dominant literature on construction management there is a large concern with tools and techniques on matters such as productivity, project delays, procurement and bidding strategies. The literature does not deal coherently with the subject, presenting a fragmented approach to the construction process. In practice, the client is rarely placed at the centre of the process; neither is the user placed at the centre. Moreover, most construction management literature seldom considers the strategic aspects of a building project as a whole. In order to address these imbalances, research has been directed towards the relationships between a building project, its context in operational and market uncertainty terms and its effect on customer (i.e. user) and client satisfaction. The key issues are presented through the use of a case study based on the refurbishment of a major office building.

Christian Lindfors contends that managing the complexity of construction, in terms of handling, controlling and directing organisations to perform as planned on projects, is key to success. Efficiency here will lead to greater certainty both for the organisation and its client. Furthermore, by eliminating unnecessary and non-value-adding activities, the client will benefit from a lower overall cost. Unravelling the complexity of the construction supply chain is fundamental to the release of better value for money. Achieving these objectives is no trivial affair and requires a degree of transparency and process efficiency that is generally not common in construction. A study is addressing the introduction of process thinking into a construction company and a structured way of making systematic representation of processes. The preliminary results point in the direction of a common platform for management information systems with a single representation and/or description as a basis for improving project efficiency.

Peter Samuelsson argues that improvement of performance is crucial if companies are to remain successful in a rapidly changing environment. This is as evident for construction as for any other industrial sector. Even so, authors claim that the construction sector is not oriented towards improving its performance. For example, practical applications of continual improvement are seldom described, and even less frequent are its applications to construction. Improvement processes introduced in a construction company are outlined and compared with current knowledge of practices elsewhere that are specifically intended to bring about improvement. The purpose is to identify key aspects of improvement processes in construction companies in the context of continual improvement. A longitudinal case study has been carried out at Skanska in order to analyse actual improvement processes, and the findings are presented.

Robert Fekete expresses the concern that the experimental studies of architects concentrating mainly on design activity may need to be complemented. Shifting focus to *what* is being designed from *how* it is being designed could contribute to design theory and consequently expand the foundations of computer-aided architectural design (CAAD). Studies of *what* is designed are best conducted by reviewing architectural theory and the writings of architects. The nature of design problems, viewed as wicked problems, supports this hypothesis. Wicked problems are typically of concern during the early stages of design, a phase in which today's CAAD systems have limited application. Questions should address the real benefits of applying CAAD early in the design process. Moreover, the concern should be with the properties that CAAD should have if it is to be of benefit during the early stages of design.

Jan Henrichsén highlights that the communication of concepts between designers and other actors in the construction process has long been a cause for concern, though much research has been done to improve understanding of the problems and practical ways forward. The use of object-oriented, computer-aided architectural design (CAAD) tools and systems has been advocated as a solution and shows promise of meaningful integration of data that can support collaborative working. Enhancement of CAAD tools to support the simulation of real world properties and processes is also under examination. Important in this regard is the use of such tools directly by the client and other project stakeholders, for instance planning and environmental authorities. A study is being undertaken to determine how CAAD can be used to simulate the effects of a proposed building in ways that relate directly to the needs of stakeholders other than designers.

Susan Bergsten reminds us that extending buildings vertically is fraught with technical and managerial problems. Inevitably, many of these types of building will be located in areas with access restrictions and other physical constraints on the movement of materials, components, operatives and equipment. The use of light-gauge steel framed systems represents a practical and cost-effective solution to the problems created by these buildings. However, materials handling and other logistical problems mean that the construction process is less than certain. The concept of 4D CAD, which has emerged from the process-engineering sector, is being increasingly considered for applications in the building sector where data regarding construction methods, resources and time are integrated with 3D design information. A major case study is being used to evaluate the potential for utilising light-gauge steel framed systems, with support from 4D CAD. Results will include a comparison of the benefits over more traditional means for design and construction management when erecting vertical extensions to existing buildings.

Finally, *Ulf Nordwall* concerns himself with the challenges faced by architects, in terms of how to design, within a reasonable, practical and financial framework, something people will like and enjoy occupying. Research is exploring the connections between good architecture, housing that is a pleasure to occupy and effective facilities management. Several questions are driving the research and these include the significance of physical attributes and architectural design to the pleasure the occupants feel, and the costs and revenues during the occupancy phase of facilities management. Others address the architectural attributes that have an impact on the rent paid for an apartment or house, and are concerned with those aspects that the occupants appreciate, as opposed to those for which the architect strives. These questions are based on sets of architectural attributes, used in the research to describe physical characteristics that influence the facility management of apartment buildings and the pleasure and comfort that occupants feel. The attributes are durability, change, renewal, material, execution and planning. They represent the tools with which the above research questions have been analysed. The fundamental purpose of the research is to gain insight into and knowledge of the relationship between architects, occupants and housing companies in terms of the design of apartment buildings, and the measurable and non-measurable attributes with respect to the occupants and their needs.

Chapter 2

Modularisation in the Customisation of Manufactured Housing

Carina Johansson

Introduction

Traditionally, construction work has been considered as a local activity, requiring local presence, local knowledge and local resources. Huemer & Östergren (2000) reveal a changed attitude within two major construction companies. The strategy for international activities of these companies has typically been characterised by treatment of each market in isolation. Pressure to respond to local requirements has forced the companies to focus on local knowledge and experience, yet they have failed to obtain economies of scale and learn from experiences within local markets. In their study, Huemer & Östergren (2000) find a changing attitude in the perception of being local by not excluding the possibility of learning from different organisational units and markets.

There is other evidence of a willingness to become more competitive in an international environment among European construction companies. The European R&D project, *FutureHome*, includes industrialists from six countries (Atkin & Wing 1999). The essence of the project involves developing know-how to create affordable, high-quality, cost-effective manufactured housing, taking account of the diversity of styles, designs and materials as well as the preferences of owners and occupiers; that is, customers who are as diverse as Europe itself. Among the specific benefits for the European construction sector arising from this project, there is the possibility of reaching a wide geographical area for innovative housing products and services.

By creating design concepts for building systems that can generate designs that are suitable for conditions in different countries and that can take advantage of advanced manufacturing methods, construction companies can become more competitive in an international setting. In order to achieve this position, they would benefit from adopting a modularisation approach to the manufacture of housing products. The research described is intended to identify the means for creating concepts and outline designs for building systems that can generate a wide range of different designs for adoption in different countries.

State-of-the-art review

Europeanisation: designing for diverse market demands

The ongoing harmonisation of technical regulations, standards and guidelines for quality will affect the whole construction sector in the European Union, but the *Europeanisation* process is continuous. Among the benefits, as a result of harmonisation, Priemus (1998) mentions promotion of interchangeability, compatibility and complementarity of components and products; the utilisation of scale effects; greater freedom of choice for customers as well as producers and contractors by access to equivalent products. *Europeanisation* will first affect the larger construction companies and it is likely there will be more mergers and acquisitions. Product development in the building materials and products (supply) sector is not tied to a specific project and is therefore likely to grow more than actual construction as a result of *Europeanisation.* Janssen (2000) points out that competition across national boundaries is hindered by institutions and regulations concerning procurement, liability, technical standards, modes of financing and institutional structures of the actors. Trans-national contracting often operates, therefore, through acquisitions or co-operation with indigenous companies. However, Janssen does not argue against the productive development of the sector that is to be expected from a higher degree of compatibility and harmonisation.

Producing for an international market includes taking diverse demands into account. Examples from the literature regarding practice in some countries suggest that initial and future design choices are not so common. Gann *et al.* (1999) compare flexibility and choice in housing in the UK, Finland and the Netherlands, highlighting some of the issues regarding current practice. The open system concept as applied in the Netherlands gives customers a greater choice regarding internal layout, but generally only for those who can pay for it. The Finnish approach to the open system concept does not usually include designing for future modification and the flexibility is concentrated in the pre-construction phase. However, in both countries there is positive attitude within government, among technical experts and researchers to develop a more customer-focused house-building sector.

Volume house builders in the UK offer few choices for the customer according to Roy & Cochrane (1999). Production ranges are in the first place defined by type of house, for example terraced house or detached house; by a small number of architectural styles, and by the number of bedrooms. Internal layout and specifications are mostly fixed for a specific product range.

The customer can choose kitchen and bathroom finishes, provided that the order is placed early enough in the building schedule. Companies with an internal architect are more willing than others to let standardisation step aside in favour of more customised solutions (Hooper & Nicol 2000). In order to be able to fulfil diverse customer requirements, Roy & Cochrane (1999) point to mass customisation and the use of product platforms from which a number of product variants can be derived.

All choices imply certain initial, running, maintenance and replacement costs. Gann *et al.* (1999) comment on the need to inform customers about choices, since housing remains a product where little information is provided on component life, running and maintenance costs and environmental issues. A study by Leather *et al.* (1998) indicates that knowledge among homeowners of costs for maintenance and repair work is generally poor; at the same time, cost is an important constraint for how and when maintenance and repair work is carried out. A component-based approach to the assembly of houses could be useful, especially if interdependencies between components can be reduced and elements separated according to different life cycles, making adaptability, maintenance, refurbishment and recycling easier (Gann *et al.* 1999).

Open building, standardisation and pre-assembly

Open building was developed as a response to the lack of customer choice and involvement during the mass housing era of the 1960s. The term *open building* describes a set of principles and techniques developed by John Habraken (Gann *et al.* 1999).

A central concept of *open building* is distinguishing between what are termed levels. The three levels of decision-making are the *tissue* level (planning and neighbourhoods), *support* level and *infill* level (Gann *et al.* 1999). The support level or base building is, according to Kendall & Teicher (2000), common to all occupants and may last for 100 years or more. In a multi-family house the support level is the load-bearing structure, common mechanical and conveyance systems, public areas and most of the outer skin. Individual occupants should not touch the support level.

Kendall & Teicher (2000) explain the infill level as all components specific to the individual dwelling unit. For a multi-family dwelling these are partitioning, kitchen and bathroom equipment, unit heating, ventilation and air conditioning system, outlets for power, communication for security, ducts and pipes and cables that individually service facilities in each unit. For a detached dwelling, *open building* distinguishes changeable interior fit-out from the more durable structure and skin.

Gann *et al.* (1999) summarise the principles of *open building* as:

- making different levels of building activity functionally and physically independent;
- maximising occupier choice by re-engineering the building, its dependencies and lead-times;
- optimising production methods and locations to deliver occupier choice at controlled levels of quality, using industrialised component systems;
- developing information systems that integrate user requirements, design, cost and production control;
- providing an effective system for delivering and installing occupiers' desired interiors on site; and
- maximising sustainability in the context of a flexible, adaptable product and accommodating the recycling of parts.

The idea of *open building* is well developed but it is not widely used in practice. Gann *et al.* (1999) mention some of the reasons for this. There is not an effective medium or large-scale industrial delivery system with supply chain linkages through design and manufacture to component and sub-system assembly. One reason for the slow commercialisation of existing infill systems may be difficulties adapting design and construction of the structure with the particular needs of a certain infill system.

Standardisation as described by Gibb (2001) is the extensive use of components, methods or processes in which there is regularity, repetition and a background of successful practice and predictability. The author argues that even though standardisation and pre-assembly are not new, their application and drivers need to be considered in the light of current technology and management practice. The two fundamental drivers for standardisation and pre-assembly are pragmatism – industry response to an urgent need combined with lack of resource and perception – and client and public reaction to a prevailing design philosophy. The conflict between maximum standardisation and flexibility has not yet been resolved. Accurate fit and interchangeability of components make standardisation work, and hence the most important area for standardisation is the interface between the components rather than the components themselves (Gibb 2001). There are three types of interfaces that are relevant to off-site fabrication, namely physical; managerial or contractual; and organisational (Gibb 1999). Physical interfaces refer to those between elements of the building, while managerial or contractual interfaces occur as a result of the subdivision of the project work content. Organisational interfaces refer to relationships between parties involved in the contract.

Modular design, product platforms and product families

Literature provides several examples of modularisation applied in practice within various types of company and industrial sectors. Baldwin & Clark

(1997) mention an example from the computer industry – IBM System/360 – that was developed using the principles of modularity in design. A family of computers that included machines of different sizes suitable for different applications was developed, all of which used the same instruction set and shared peripherals. IBM succeeded greatly by making the new system compatible with existing software. The automobile industry and financial services provide other examples (Baldwin & Clark 1997). The components of an automobile are manufactured at different places and brought together for final assembly. It is possible because design of each part is precisely defined. Services are also being modularised, for example in the financial service industry. Financial services are intangible and the *science of finance* is well developed, which make its products relatively easy to define and separate. As a result, providers do not have to take responsibility for all aspects of delivering a certain financial service.

Modularity is referred to as the division of products into smaller building blocks or modules – see Fig. 2.1. A modular system is a set of modules with which product variety can be created. Hence, common elements and interfaces are the platform and all elements and interfaces that together make up the modular system. The product variants constitute the product family (Blackenfelt & Stake 1998). The platform is a set of sub-systems and interfaces that form a common structure from which a stream of derivative products can be efficiently developed and produced (Meyer & Lehnerd 1997).

This description is focused on the actual product. Meyer & Lehnerd (1997) discuss a broader view of the product platform concept where the product platform consists of building blocks. These are divided into four categories.

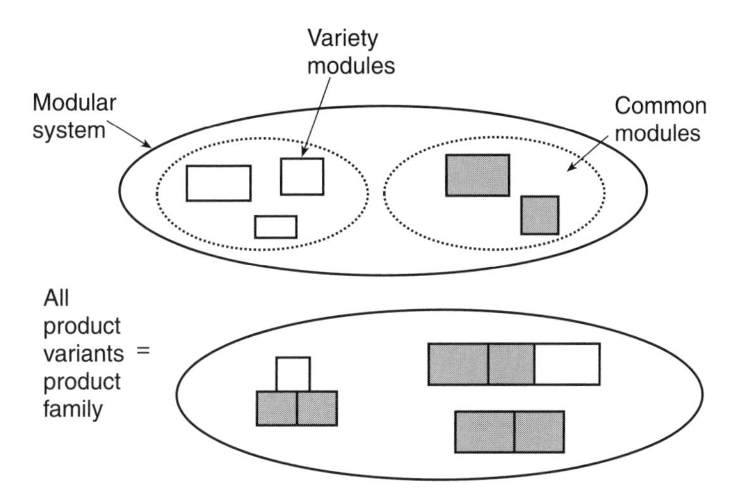

Fig. 2.1 Modular system and product family (Source: Stake & Blackenfelt 1998).

(1) Insights into the minds and needs of customers and the processes of customer research and competitive research that uncover and validate those insights.
(2) Product technologies in components, materials, sub-system interfaces and development tools.
(3) Manufacturing processes and technologies that make it possible for the product to meet competitive requirements for cost, volume and quality.
(4) Organisational capabilities, which include infrastructures for distribution and customer support, as well as information systems for control and market feedback.

Stake (2000) suggests that a platform can be considered as something that is common for a range of products and distinguishes between two ways of considering the product platform:

- The commonality-based viewpoint, which regards the common elements of the product family to be the product platform.
- The resource-based viewpoint, which regards all the common resources necessary to build the product family to be the product platform.

Product architecture and modularity

Quick and cost-effective creation of derivative products can be achieved when interfaces between sub-systems of products and between products and users are defined (Meyer & Lehnerd 1997). The combination of sub-systems and interfaces define the product architecture and the goal is to make the architecture common across many products. This is the essence of modularity.

A modular architecture is according to Ulrich & Eppinger (2000) a product architecture in which each physical *chunk* implements a specific set of functional elements and has well-defined interactions with the other chunks. The opposite is an integral architecture in which functional elements are implemented using more than one chunk and/or a single chunk implements many functional elements. Interactions between the chunks are ill-defined. Products are seldom strictly modular or strictly integral. Ulrich & Eppinger (2000) point out that the product architecture has implications for product change, product variety, component standardisation, product performance, manufacturability and product development management.

Rising minimum levels of acceptability for product development performance is a reality in many industries (Robertson & Ulrich 1998). The authors point out that it is no longer possible to dominate large markets by developing one product at a time and that, increasingly, good product development means good platform development.

Modularity is often discussed in the context of product variety. Baldwin & Clark (1997) mention the explosion of product variants and point out that modularity can reduce the problems associated with product variety and manufacturing. This is achieved by increasing the degree of commonality between product variants at the same time as allowing the customer to mix and match elements to suit his or her preferences. Modularity provides the assembler with flexibility by allowing for the manufacturing process to be delegated to many different suppliers, both internally and externally, as pointed out by Baldwin & Clark (1997).

Erixon (1998) mentions some of the benefits of modular products, based on case studies of different companies.

- A modular design is a robust basis for product renewal and concurrent development of the production system.
- Short feedback links for failure reports can be secured if the modules are tested before delivery to the main line.
- Increased modularity of a product leads to positive effects on the total flow of information and materials from development and purchasing to storage and delivery.
- Combining a modular design with product planning simplifies the product development process and planning of corresponding production system changes.
- Increased modularity leads to a reduction of throughput times. An early fixing of interfaces between modules allows for product development activities to take place at the same time, but separately for each module.

Considering all the advantages with modular products, how is it that not all products are modular? Baldwin & Clark (1997) point out that it is more difficult to design a modular system than a comparable interconnected system. The designer's knowledge and experience play an important role in creating feasible modular concepts, as it would otherwise be easy to create concepts that are not workable from a technical or spatial perspective (Erixon 1998). Robertson & Ulrich (1998) mention two common problems in companies working towards the creation of product platforms.

The first problem observed is that organisational forces prevent the ability to balance commonality and distinctiveness when only one perspective dominates, for example costs for creating product variation. The other problem concerns the dilemma of a balanced team handling platform planning. The team can go too deeply into details, resulting in the organisation giving up or the product itself lacking character and integrity. Robertson & Ulrich emphasise that top management's participation is needed because making good platform decisions requires making complex trade-offs in different business areas.

Dividing products into modules

Several authors have developed methods and tools for modularisation. These can generally be divided into four categories: methods spanning from market activities to detail design, methods for starting with a product specification and ending with a defined product architecture, guidelines for re-use and commonality, and guidelines for creation of product platforms (Blackenfelt & Stake 1998). A few methods are briefly described here.

Meyer & Lehnerd (1997) describe a method which is included in the category of providing a guideline for the creation of product families. The method is referred to as composite design, and includes six steps. These are:

(1) establish the goals of the system in terms of performance and price;
(2) classify and analyse the sub-systems of their own design and those of the competitors;
(3) measure the design's complexity and those competitors' products;
(4) index the design and those of direct competitors against standard baselines of function and cost;
(5) build in degrees of freedom to accommodate product line expansion; and
(6) integrate manufacturing processes to achieve lowest cost.

Since it is relatively easy to create many product variants, it is important to provide only the variants that customers demand. Meyer & Lehnerd (1997) provide a tool, the platform market grid, to support evaluation of the platform against different market segments.

Erixon (1998) has developed the Modular Functional Deployment (MFD) method, which is included in the category of methods spanning from market activities to detail design. The method includes five steps:

(1) clarify customer requirements;
(2) select technical solutions;
(3) generate concepts;
(4) evaluate concepts; and
(5) improve each module.

The core part of the MFD method is step three, where the module concepts are generated by evaluating technical solutions against *so-called* module drivers. The latter are the primary reasons in opting for modularity. Erixon (1998) has identified twelve generic module drivers from case studies. In addition to these there may also be company-specific module drivers.

Development and design

- Carry-over: part of a product or sub-system that can be re-used in a future product generation.
- Technological evolution: a part or a sub-system that is likely to go through a technological shift during the lifetime of the product family. This can be caused by, for example, radically changing customer demands.
- Planned design changes: the part or sub-system is scheduled to go through a change according to an internally-decided plan.

Variance

- Technical specification: the part or sub-system varies in terms of function or performance between the products in the family.
- Styling: the part or sub-system varies in terms of colour and shape between the products in the product family.

Production

- Common unit: the part or sub-system can be used across the whole product family.
- Process and/or organisation re-use: the part or sub-system suits a certain process or has suitable work content for a group.

Quality

- Separate testing: the part or sub-system should be tested separately.

Purchasing

- Supplier available*:* the part or sub-system may be outsourced.

After sales

- Service and maintenance: the part or sub-system needs to be easily serviced and maintained during the life of the product.
- Upgrading: the unit may be replaced for another part with different function or performance.
- Recycling: environmentally hostile or easily recyclable materials can be kept apart to make recycling and disposal easier when disassembling the part or sub-system.

Stake (2000) illustrates a simple strategic company approach based on a module driver profile. Three approaches, which are related to specific module drivers, are suggested.

(1) Product – the company focuses on either having superior products compared with its competitors or incorporating new technology sooner. The related module drivers are carry over, different specification, technology push and planned design changes.
(2) Process – the company focuses on the manufacturing process to achieve economical manufacturing by, for example, economies of scale. The related module drivers are carry over, common unit, process/organisation, separate testing, strategic supplier and recycling.
(3) Customer – the company focuses on the customer by providing product variety and by emphasising after-sales aspects. The related module drivers are different specification, styling, service/maintenance, upgrading and recycling. Companies usually combine the approaches and some module drivers are related to several strategies.

The Design Structure Matrix (DSM) – also known as The Dependency Structure Matrix, the Problem Solving Matrix and the Design Precedence Matrix – is a system analysis tool that can be used to represent a complex system. It achieves this by capturing the interactions, interdependencies and interfaces between sub-systems and modules. DSM is also a project management tool, which provides a project representation that allows for feedback and cyclic behaviour. The DSM method belongs to a category that starts with product specification and ends with a defined architecture.

Lanner & Malmqvist (1996) suggest an approach that considers technical and economic aspects in product architecture design. The developed method combines two methods for product structuring, one that focuses on the technical aspects and another that considers economic factors, effectively an interaction matrix extended by the score of the economic factors.

Yu *et al.* (1999) propose a customer-need basis for defining the architecture of a portfolio of products. A market population is assessed to establish target values for product features at a specific time and represented as a probability distribution. The desired values are also assessed for a sample of representative customers over time and represented as a probability distribution. By comparing the two distributions for important customer needs it is possible to pinpoint the type of product architecture a given market population desires.

Whole life costing

Whole life costing is a proven means for measuring how design options affect the initial and operating costs of buildings. It can be defined as:

> 'the present value of the total cost of that asset over its operating life including capital costs, occupation costs, operating costs and the cost or benefit of the eventual disposal of the asset at the end of its life'. (Hoar & Norman 1990)

The main components of whole life cost are present or future capital sums, recurring costs (i.e. running and maintenance costs), and a sinking fund to repay the capital at the end of the life of the asset (Ferry *et al.* 1999). To calculate whole life costs, all costs relevant to a project are added to estimate the total cost. Future costs are discounted to present-day values.

Whole life costing can be used as a decision support tool during the design phase. In order to perform a whole life cost analysis the adequacy of the predicted amount and timing of future resources are core issues. Whole life costing is particularly useful for option selection, for example determining whether a higher initial cost is justified by a reduction in future costs or identifying whether or not a proposed change is cost-effective against the *do-nothing* alternative.

Research project

Scope

This research addresses aspects of internationalisation and customisation within the house-building sector in Europe, with the aim of increasing understanding of how to create concepts for modular building systems that can generate housing designs adaptable to conditions in different countries.

A building system can be modularised in many different ways, depending on the goals for modularisation and the module drivers that are to be emphasised. A systematic approach to support product modularisation is adopted to increase understanding of how to create modular building system concepts, adaptable for international markets having differing local conditions; considering product variance requirements that allow for maintenance, replacement and upgrading to be undertaken economically over the building life cycle.

The objectives are three-fold.

(1) To describe a framework for a decision process for designing a building system that supports the achievement of defined targets, based on customer and company requirements. A modularisation approach to the design will be adopted.
(2) To test and evaluate the framework for a decision process in a detailed case study within a company.
(3) To refine the framework and make recommendations for implementation.

The level of detail aimed for in the framework described for a decision process is intended to allow for identification of conceptual modules, i.e. groups of design parameters that correspond to the reason for being a module: this is known as a module driver.

The research is limited primarily to the module drivers of *product variance* and *after sales*, because these are two areas that are judged to be among the weakest, and for which improvement could yield worthwhile benefits for the sector, as well as superior products for customers. As required, to allow the generated framework to produce feasible solutions, other module drivers are taken into consideration.

Research methodology

The research problem addresses the issue of creating concepts for modular building systems to generate housing designs that can be adapted to conditions in different countries. The focus is on how to produce design concepts that can be adapted according to product variance requirements, and that allow for maintenance, replacement and upgrading to be done economically over the building life cycle. Potential methods for this research problem can be found in literature on product design and development, the *open building* approach and product modularisation.

Ulrich & Eppinger (2000) describe a generic product development process. Particular parts of the generic process will differ according to a company's specific context. The process is most likely to be used in a market-pull situation, i.e., a company starts product development with a market opportunity and uses available required technologies to satisfy market needs. In addition to this situation several variants are common and correspond to technology-push products, platform products, process intensive products and customised products. The process description includes some specific aspects on issues applicable to the development of product families.

Literature on *open building* provides insight into the relevance of the principles and approaches adopted in this research. An essential part of modularity, product structuring, has several ideas in common with *open building*. This is exemplified by the *open building* approaches towards separating the support and infill levels and co-ordinating and separating the sub-systems. The distinction of levels and co-ordinating of sub-systems share with modularisation the notion of defined interfaces to obtain a product structure that allows for interchangeability of components and facilitates maintenance. Sarja (1997) describes what he terms *open industrialisation* – the application of modern systematised methods of design, planning and control as well as mechanised and automated manufacturing processes. Openness refers to the ability to assemble products from alternative suppliers into the building and to exchange information between actors in the construction process. Sarja describes a method for integrated life-cycle design, and points out that a modularised systematic approach helps to rationalise design.

Existing methods for product modularisation can generally be divided into four categories (Blackenfelt & Stake 1998):

- methods spanning from market activities to detail design;
- methods for starting with a product specification and ending with a defined product architecture;
- guidelines for re-use and commonality; and
- guidelines for creation of product platforms.

By combining and adapting existing methods, a framework for a decision process for designing modular building systems, that supports the achievement of defined targets based on customer and company requirements, can be generated.

Research results and industrial impact

Tentative results

The literature points to the need for increased customer influence in dwelling design and to the need for informing customers about the implications on life-cycle costs for certain choices. International competitiveness is becoming increasingly interesting in the house-building sector. Producing for an international market includes satisfying diverse needs. Modularisation is one way of handling product variety.

There are various methods for creating modular products that focus on different aspects. Some focus on technical issues, some focus on reusability of parts across the product family and others include the company's strategic concerns. The chosen research approach is, therefore, to combine and adapt existing methods to deal with the research problem and, in particular, to meet customer requirements effectively.

The description of a decision process for modularisation, that takes proper account of customer requirements, can result in a framework to support the achievement of an internationalised product platform. This would be capable of producing dwellings with high customer acceptability, while supporting increased industrialised building. There are clear benefits to housing manufacturers, not least in providing economies of scale and the means for satisfying greater demand.

Implementation and exploitation

So far, the *Europeanisation* of the construction sector has been seen mainly in the context of large companies involved in trans-national projects, such as the Channel tunnel and Öresund bridge, and those with associated companies and subsidiaries across Europe. The large majority of European construction companies – the small and medium-sized – still act mostly locally. These companies are affected by differences in regulatory requirements and various kinds of trade barriers that inhibit international trade in their products.

Harmonisation of technical regulations and standards will lead to increased movement of construction services and products across the European Union. A systematic modularisation approach can be of benefit to companies that operate internationally, as a support for issues such as coordination of manufacturing and purchasing of components across its markets. However, there will always be a question of trade-offs between different aspects of manufacturing, assembly, standardisation and the number of choices the customer can be offered. The framework, which this research intends to create, will provide a guide for addressing the issue of finding balance between standardisation and flexibility at the same time as it will allow for generating a building system that can be manufactured, assembled and used efficiently.

Conclusions

The house-building sector is facing the challenge of meeting the requirements of the customers in an international market. Literature provides examples of an ambition among construction companies to become competitive in international markets and benefit from economies of scale. There are also examples in the literature of the requirement for house builders to provide more choice in housing and to inform owners and occupiers properly of the implications of certain choices on life-cycle costs.

Open building is a concept that provides the owner and occupier with choice initially, as well as at later stages of the building life cycle. *Open building* as applied in practice has so far not provided a wide range of people with choice, due to factors such as the lack of a medium- or large-scale industrial delivery system that can contribute to economies of scale, thereby reducing end-user costs. For housing producers aiming at international markets, creating design concepts that can be adaptable to local conditions requires many factors to be taken into account. They must allow for, among other things, building traditions and regulations, owner's preferences, and the whole life costing of the building system and its components.

In order to improve knowledge on how to create a modular building system that can generate dwellings capable of being built in different countries, product modularisation methods are adapted and applied. Such modified methods aim at improving the process of generating internationalised building systems and informing the customer of the implications of choices upon life-cycle costs.

References

Atkin, B.L. & Wing, R.D. (1999) FutureHome – Manufactured Housing for Europe, *16th IAARC International Symposium on Automation and Robotics in Construction*, Madrid, September 1999, pp. 573–8.

Baldwin, C. & Clark, K. (1997) Managing in an age of modularity. In: *Markets of One – Creating Customer-Unique Value through Mass Customisation (2000)* (eds Gilmore, J. & Pine II, J.), Boston: Harvard Business School Press, pp. 35–52.

Blackenfelt, M. & Stake, R.B. (1998) Modularity in the context of product structuring – a survey, *Proceedings of NordDesign Seminar*, 26–28 August, Stockholm: Royal Institute of Technology, pp. 157–66.

Erixon, G. (1998) *Modular Function Deployment*, Doctoral thesis, Stockholm: Royal Institute of Technology.

Ferry, D., Brandon, P. & Ferry, J. (1999) *Cost Planning of Buildings*. Oxford: Blackwell Science.

Gann, D., Biffin, M., Connaughton, J., *et al.* (1999) *Flexibility and Choice in Housing*. Bristol: The Policy Press.

Gibb, A. (1999) *Off-site Fabrication: Prefabrication, Pre-assembly and Modularisation*. Caithness: Whittles Publishing.

Gibb, A. (2001) Standardization and pre-assembly – distinguishing myth from reality using case study research, *Construction Management and Economics*, **19**, pp. 307–15.

Hoar, D. & Norman, G. (1990) Life Cycle Cost Management. In: *Quantity Surveying Techniques – New Directions* (ed. Brandon, P.S.), Oxford: BSP Professional Books, pp. 139–68.

Hooper, A. & Nicol, C. (2000) Design practice and volume production in speculative housebuilding, *Construction Management and Economics*, **18**, pp. 295–310.

Huemer, L. & Östergren, K. (2000) Strategic change and organisational learning in two Swedish construction firms, *Construction Management and Economics*, **18**, pp. 635–42.

Janssen, J. (2000) The European construction industry and its competitiveness: a construct of the European Commission, *Construction Management and Economics*, **18**, pp. 711–20.

Kendall, S. & Teicher, J. (2000) *Residential open building*, London: E & F.N. Spon.

Lanner, P. & Malmqvist, J. (1996) An Approach Towards Considering Technical and Economic Aspects in Product Architecture Design. *2nd WDK workshop on Product Structuring*, Delft, The Netherlands, 3–4 June, Delft: Delft University of Technology, pp. 173–80.

Leather, P., Littlewood, M. & Munro, M. (1998) *Make Do and Mend: Explaining Home-Owners' Approaches to Repair and Maintenance*, Bristol: The Policy Press.

Meyer, M. & Lehnerd, A. (1997) *The power of product platforms,* New York: The Free Press.

Priemus, H. (1998) The impact of European integration on the construction industry. In: *European Integration and Housing Policy,* (eds Kleinman, M., Matznetter, W. & Stephens, M.), London: Routledge, pp. 77–93.

Robertson, D. & Ulrich, K. (1998) *Planning for Product Platforms.* Sloan Management Review, Summer, pp. 19–38.

Roy, R. & Cochrane, S.P. (1999) Development of a customer focused strategy in speculative house building, *Construction Management and Economics,* **17**, pp. 777–87.

Sarja, A. (1997) Research and development towards sustainable open industrialization. In: Building Function, Environment and Technology, *Proceedings of the Scandinavia–Japan Seminar on Future Design and Construction in Housing,* 6–8 October, Stockholm: Royal Institute of Technology, pp. 71–8.

Stake, R. (2000) *On conceptual development of modular products – Development of supporting tools for the modularisation process,* Doctoral thesis, Stockholm: Royal Institute of Technology.

Stake, R.B. & Blackenfelt, M. (1998) Modularity in use – experiences from five companies. *4th WDK Workshop on Product Structuring,* 22–23 October, Delft: Delft University of Technology.

Ulrich, K. & Eppinger, S. (2000) *Product Design and Development,* Boston: McGraw-Hill.

Yu, J., Gonzales-Zugasti, J. & Otto, K. (1999) Product Architecture Definition Based Upon Customer Demands, *Mechanical Design,* **123**, *3*, pp. 329–35.

Chapter 3

Application of Integrated Life Cycle Design to Housing

Mats Öberg

Introduction

The construction and operation of buildings has a tremendous impact on the economy and environment. In the countries of the EU, construction accounts for, on average, 11% of GDP. At the European level, the construction sector employs 30 million workers, is responsible for 40% of energy and materials consumption and generates a similar level of waste (Sjöström 2000). A substantial part thereof is related to the user phase, for instance 85% of the energy use (Adalberth 2000). In April 2001, the European Commission adopted a proposal for a Directive on energy efficiency in buildings prompted by the concern that as much as 40% of the CO_2 emissions in Europe are attributable to existing buildings. For the purpose of sustainability it is therefore crucial that the built environment is carefully designed and that life cycle aspects are properly assessed.

The building sector is faced by the challenges of increasing productivity and safeguarding or enhancing the long-term quality of its products. According to several investigations and commissions such as the *M4I, Movement for Innovation,** in the UK, and *Byggkostnadsdelegationen* (BKD 2000) in Sweden, a reduction in cost of 10–20% and production time of 10% should be achieved for buildings to ensure the affordability of housing. Increased environmental consideration is adding other requirements to buildings such as decreased energy use and improved indoor climate.

Rationalisation of the production process and development of the intended building product need to be based on a holistic life cycle perspective. Integrated life cycle design (ILCD) is intended for this purpose and is thus a tool for supporting the pursuit of sustainable development. When considering the overall impact of the building sector, it is a question of when, rather than if, it

*M4I is a cross-sector initiative aimed, *inter alia*, at the widespread implementation of improved processes and product delivery.

is generally introduced. The prime obstacle is possibly the fragmented construction process itself, where the designer and/or producer lack interest in or knowledge about the occupancy (or use) phase. For the frontrunners, ILCD provides competitive advantage. For the facilities owner and user the advantages are obvious, as they are for those engaged in design, build and operate contracts. Even those involved in more traditional forms of building procurement can gain from the systemisation that follows from ILCD.

The *open building* concept is a modular, systematic approach for adopting and organising the building, that answers to the call for design to be based on life cycle principles. Experience from successful application of *open building* in, for example, Finland, Japan and the Netherlands should be further explored and is, indeed, examined in some detail in Johansson's Chapter 2.

State-of-the-art review

Integrated life cycle design (ILCD) involves some methodological factors that are not generally found in practice within the construction sector today, specifically life cycle costing, life cycle assessment and multiple criteria decision-making. With regard to practical applications, it is presumed that the concept of *open building* fits the principles of life cycle design. These two topics, ILCD and *open building*, are discussed further below.

Integrated life cycle design methods

Life cycle optimisation of products calls for the application of tools for whole life appraisal and methods to control and organise the decision-making towards multiple criteria. *Integrated life cycle design* or *holistic product development* can be furnished by incorporating three related tools into the traditional design procedure: *quality function deployment* (QFD), *life cycle assessment* (LCA) *and life cycle costing* (LCC). Sarja (1998, 2000) has proposed a framework for ILCD for buildings. His framework outlines the scope of ILCD and identifies relevant methods and tools – see Fig. 3.1.

Multiple criteria decision-making can be aided by QFD, which is a method for organising, collecting, interpreting and assessing the client's needs with regard to functionality. QFD is used to support product optimisation in the design phase and has been successfully applied in manufacturing industry since the early 1980s. Akao (1990) defines QFD as a method for a) developing a design quality aimed at satisfying the customer, and b) translating the consumers' demands into design targets and major quality assurance points to be used throughout the production stage. QFD is thus a systematic way for tuning the product's features to the client's requirements and for documenting

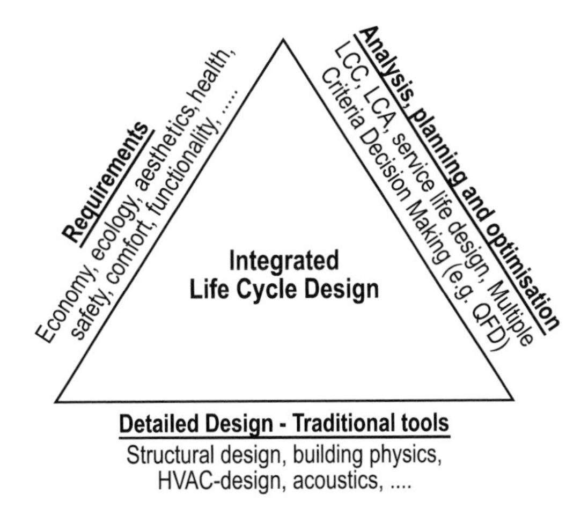

Fig. 3.1 Framework for integrated design of buildings (after Sarja 1998).

the decisions taken in the design process. It is organised as a set of matrices that address and define information such as:

(1) clients' requirements and needs;
(2) product characteristics that are required to match 1;
(3) the relationship between 1 and 2;
(4) results from analysis performed regarding alternative solutions;
(5) opposition between characteristics; and
(6) target values for the required characteristics.

QFD has been tried within the construction sector on an experimental basis, but has not yet achieved any large impact in practice. The International Energy Agency (IEA) has applied QFD to support the design of eco-efficient buildings from basic functional criteria: life cycle costs, resource use, architectural quality, indoor air quality, functionality and environmental loading. Nieminen & Houvila (2000) report some pilot cases where QFD was used with promising results, both for the overall design of new buildings and for office refurbishments.

QFD may lead to extensive paperwork with several interacting matrices at various levels of detail. This may discourage or prevent its introduction within the construction sector. Gargione (1999) mentions the following difficulties: substantial increase in time expended by the project management team, difficulties with working with large QFD matrices and in processing the information within them. There is, however, no obvious short cut to rational and systematic design decisions and the challenge will be to adapt the concept to the needs of the construction sector. Increased industrialisation of the sector with more continuous production will probably work in favour of methods like QFD.

One of the specific tools applied within ILCD is life cycle costing (LCC), which is used for identifying and quantifying consequences regarding costs related to a product during its entire life cycle. The term LCC has been defined by Kirk & Dell'Isola (1995) as:

> 'an economic assessment of investment alternatives that considers all significant costs of ownership discounted over the lifetime of a product.'

Some authors, for example Gluch (2000), refer to the US Army in the early 1960s for the first application of LCC. It could, however, be argued that the general concept and calculation procedures can be traced back to the early 1930s when methods such as *simple payback, net present value* and *internal rate of return* were introduced within manufacturing industry for calculating the economic results of investments. In those applications the profit is also taken into account (Dale 1993). During the 1950s, LCC was introduced into the construction sector by Stone at the Building Research Establishment in the UK under the name of *cost-in-use* (Ashworth 1993). Stone used present values and annual costs to assess the cash flows throughout the lifetime of a building. LCC replaced the term *cost-in-use* in the UK largely through the work of Flanagan & Norman (1983) who, together with Meadows and Robinson, also published the oft-cited handbook *Life Cycle Costing: Theory and Practice* (Flanagan *et al.* 1989).

The British Standards Institution released BS 3811 in 1974, describing the life cycle phases from design to replacement, that fits well with the concepts of LCC (and LCA). Further development of the method continued in the UK during the 1980s promoted by The Royal Institution of Chartered Surveyors (RICS 1980; 1986). There is also a Norwegian Standard, NS 3454, *Life cycle costs for building and civil engineering work – Principles and classification* published by the Norwegian Council for Building Standardization.

In Sweden, public buildings have been analysed using LCC techniques since the 1970s (Arthursson & Sandesten 1981). This has applied primarily to investments regarding refurbishments and technical equipment. Sterner (2000) describes the experience of whole life costing (LCC) in a case study comprising 54 Swedish users. The two top-ranked constraints regarding use of the method were found to be experience and input data. In terms of reliability, the respondents believed that LCC estimates were either mainly or in some way correct in 80% of cases.

Rutter & Wyatt (2000) argue that LCC is seen only as a tool for initial decisions regarding design options and that there should be a firmer link between original intentions and subsequent implementation. To illustrate the point, the term LCC should be replaced by *whole life costing*. However, the underlying argument is rather weak, as there is nothing that prevents the inclusion of the implementation phase in LCC.

LCA is a methodological parallel to LCC aimed at quantifying environmental aspects. Among the early applications of LCA were the Midwest Research Institute and the Coca Cola Company that in a study from 1969 compared environmental load and resource use for different types of beverage containers. In the 1970s, as a result of the oil crisis, focus was on production and the use of energy. The development of LCA was closely related to a limited number of persons and institutions such as Boustead (UK), Franklin (US) and Sundström (Sweden).

The next phase in the development of LCA occurred during the 1980s when it was acknowledged that physical locations for the deposition of waste material were restricted. This stimulated an interest in recycling and reuse. By the early 1990s, LCA gradually appeared more generally and the method became standardised within the ISO 14040 series. SETAC, the Society of Environmental Toxicology and Chemistry, has coordinated this work and developed a code of practice (SETAC 1993).

A large number of LCA studies regarding building products and systems and also entire buildings have been undertaken. There are many critical aspects to consider; for instance, cut-off criteria, allocation, data quality, functionality and weighting methods. These issues can usually be treated appropriately when working with the application of LCA internally within an organisation for product or process improvement; the method is a good tool for this purpose. When comparing products of different character, for example steel, concrete and timber, any conclusion must be very carefully judged and so the requirement for transparency is very high. These problems are apparent in a report by Andersson (1998), who examined four different LCA studies, initiated by organisations using competing materials for producing waste water pipes. The results were divergent and the lesson to be learned is to be critical of any statement ranking the environmental impact of significantly different products.

The linking instrument between LCA and life cycle design is primarily the EPD (Environmental Product Declarations). Type III EPD, which are third party assessed declarations according to ISO 14025, are based on the inventory part, *life cycle inventory,* of an LCA. Type II declarations – self-declarations according to ISO 14021 – are also often based on a life cycle inventory. A complete and up-to-date set of EPDs, preferably of the type III, for the assurance of reliability of building products and systems, is a key enabler for integrated life cycle design (ILCD).

Open building

Open building supports strategies for programming, design, production planning and facility management in organising buildings physically into different layers. These extend from the entire urban fabric (tissue), over the building frame (support) to fit-outs of components and systems (infill) and finally furnishings, fixtures and equipment (FF&E). The key features are design for flexibility and appropriate longevity and disentanglement of systems. The modular approach of *open building* allows for systematic optimisation of the target service life, life cycle cost and environmental aspects of different parts of a building (Sarja 2000). One important idea about *open building* is what can be referred to as *mass-customisation*. During the 1960s, the effectiveness of the production of multi-family dwelling buildings in Europe rose dramatically. This development was, however, founded on an extreme production-orientation with large-scale projects and high repetitiveness. The attitude today towards this kind of built environment is generally negative. *Open building* promises similar productivity but irrespective of project size, and with a high degree of architectural freedom. *Open building* should fit in well to the concepts of lean and agile production as well as supply chain management, which are regarded as critical to successful production within manufacturing industry. These concepts are gradually finding their way into the construction sector. Agile production is more open to customisation than lean production and should thus be preferred to avoid problems of over-uniformity.

The initiator of *open building* was John Habraken of the Dutch Foundation for Architects' Research (SAR), which published two important reports: SAR 65 (*Basic method for designing residential supports without predetermining the size or layout of dwellings*) and SAR 73 (*A methodology for the design of urban tissues*). SAR also introduced the modular system now generally employed by the construction sector. By the mid 1970s, Habraken moved to MIT in Boston to head the architectural department. The most significant publications regarding *open building* from Habraken are: *Supports: An Alternative to Mass Housing* (1972) and *The Grunsfeld Variations: A report on the Thematic Development of an Urban Tissue* (1981).

A current example of *open building* from Japan (Yashiro 2000) is a system for up to seven-storey, multi-family dwelling buildings. The building frame (support) consisting of concrete columns and beams is designed for 200 years, which is extremely long for Japanese conditions, and was in this case considered as reasonable in the context of assuring sustainability requirements. The spatial composition of the interior is highly adaptable and even certain concrete floor slabs may be removed. All technical service systems are easily accessible for exchange or repair.

Kendall (2000) has presented a comprehensive overview on the theory and the practical application of residential *open building*. The examples indicate that there is a critical and sometimes rather diffuse line between *mass-customisation* and *mass-production*, and that care must be taken to avoid repeating the over-industrialisation typical for the large-scale residential projects of the 1960s. Many architects and city planners are profoundly sceptical of construction-kit systems that are geared primarily to satisfying the requirements of the production phase. Flexibility in design in regard to planning and aesthetics and between and within projects is crucial to the implementation of *open building*.

Research project

Project description and objectives

Integrated life cycle design (ILCD) is being applied to concrete multi-family dwelling buildings to verify how long-term characteristics such as economy, function and ecology can be predicted and optimised from the priorities given for a specific project. By doing this it is expected that the overall functional quality of the building over the entire life cycle can be enhanced.

The ILCD of a building involves a large number of aspects or parameters that are more or less interrelated. For example, minimised use of energy is beneficial with regard to cost, but may affect the indoor air in a negative way to the extent that energy is reduced by decreased ventilation. Several functional aspects can be expressed by classes, such as sound insulation or insurance (fire safety), with a minimum standard stated in codes. Some aspects, for example stability, are strictly defined by codes. Other characteristics such as flexibility or aesthetics are more difficult to define. On the other hand, external environmental (ecological) aspects and economy can be regarded as functions of the particular design chosen to fulfil the fixed/classified requirements. Thus the function of ILCD is, first, to set all relevant quality requirements and, second, to optimise the design with regard to economy and ecology. Ranking of different aspects should be up to the individual project. Special tools, such as QFD, support the process of multiple criteria decision-making and the applicability of such methods in building design will be particularly assessed within the research project. A common currency of ecology and economy will be used to make those aspects comparable and will be examined. Energy use can, to a certain extent, quantify ecology. Saari (2001) uses this simplification when assessing total buildings. It is acknowledged that energy use does not cover the entire environmental field. Even so, recent investigations based on LCA confirm that energy use during the user phase is the most important environmental aspect regarding the built environment (BK 2001). A second reason for this simplification is that complete, robust and reliable environmental data on building products is only available for a limited number of products and materials, making any detailed LCA for complete buildings problematic.

The introduction of life cycle design into the routine of design is not a trivial matter. It is anticipated that new integrated IT solutions, including CAD and quantity surveying, can be expanded with life cycle design tools and that both data and systems could also be utilised in a logical way within the facilities management phase. Feedback of information to the design team would then be secured, so that the oft-debated link between designer, supplier, producer and user will be reinforced. Thus, the inconvenience of changing and introducing new design routines is compensated by improvements of the entire construction process. This issue will be examined within a full-scale demonstration project.

Research methodology

Methods for ILCD are under review and, where appropriate, refined before being verified and then finally applied in a full-scale project. Prediction models and data covering costs and energy use are being collected. A select set of tools has been calibrated with actual figures from field studies of modern multi-family dwellings. Deficiencies in the models and data have been defined and improved, where necessary. Methods of multiple criteria decision-making such as QFD have also been reviewed and the possibility of integrating an assessment of life cycle aspects in an effective way within the regular design process is now being explored. Alternative technical solutions will be theoretically examined by using the tools referred to above and possible improvements with regard to LCC and energy use will be quantified.

Adaptations of cast in situ or prefabricated concrete structures to the principles of *open building* will be explored and the project will be concluded with a full-scale demonstrator that will be constructed with the aim of meeting the functional quality required by the client. This will reflect a high degree of flexibility regarding both production and use and then optimised with regard to LCC and energy use.

Case study: research results and impact

Quantification of results

The research has shown that LCC can be applied in the design phase to assess total costs for alternative design solutions with a reasonable degree of accuracy. The ratio of production to operational costs over the life cycle for multi-family dwellings is roughly 45:55 (Johansson & Öberg 2001). According to Griffin (1993), 50–80% of the total cost will occur during use. Furthermore, a large part of the total present value of costs is dependent on the design of the building, as shown in Fig. 3.2. This indicates that there is a large potential for cost saving from careful design. The most significant operating cost categories, energy and periodic maintenance, have also substantial environmental impacts for which life cycle optimisation is beneficial.

It should be stressed that over recent decades roughly half of all investment in the residential building market in the EU (with small regional and annual variations) is attributable to reconstruction (SCB 2000). Such a statistic surely underscores the importance of adopting a life cycle perspective.

Both sufficiently accurate tools for the prediction of energy use in buildings and detailed information on the need for periodical maintenance for different materials and systems are available for making LCC assessments. In Table 3.1, examples of periodic maintenance data are presented and, in Fig. 3.3, a life cycle cost comparison of three different façade designs illustrates the importance of considering the operational phase. The timber façade has the lowest production cost, but if life cycle costs are considered the concrete alternative is more economical. Table 3.1 could be expanded with environmental data for each activity. Note that there are often secondary effects of a certain design choice, such as insurance or heating costs, which need to be taken into account and these are shown in Fig. 3.3.

Preliminary results from desktop studies show that the energy use in residential buildings can be cut by 3% using simple measures. This corresponds to an annual release of 300 000 tonnes of CO_2 for Sweden's house building sector. Given a socio-economic value for CO_2 emissions of 0.16 €/kg (Vägverket 1999) this is equivalent to €48 million.

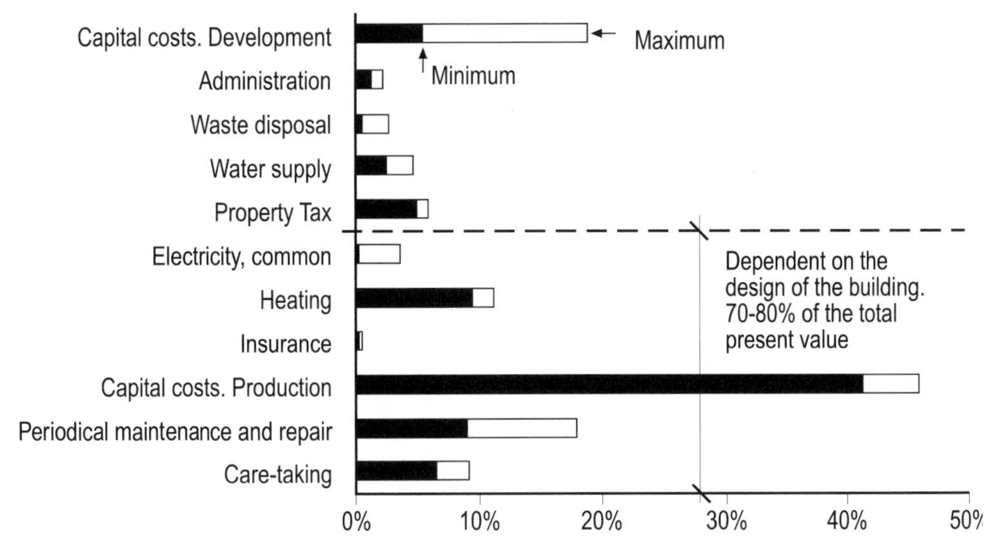

Fig. 3.2 Present value (relative) of life cycle costs for four different residential buildings according to Johansson & Öberg (2001).

Table 3.1 Periodic maintenance costs (*according to SABO (1999) and **according to Johansson & Öberg (2001)).

Façade type	Maintenance activity	Interval* (years)	Cost per activity* (€/m²)	Present value** (€/m²)	Annual cost** (€/m²)
Brick	Overhaul joints	40	135	50	1.60
Timber panel	Repaint	7–10	65	230	7.40
Timber panel	Exchange	35	345	145	4.70
Concrete panel	Repaint	14–20	60	82	2.70

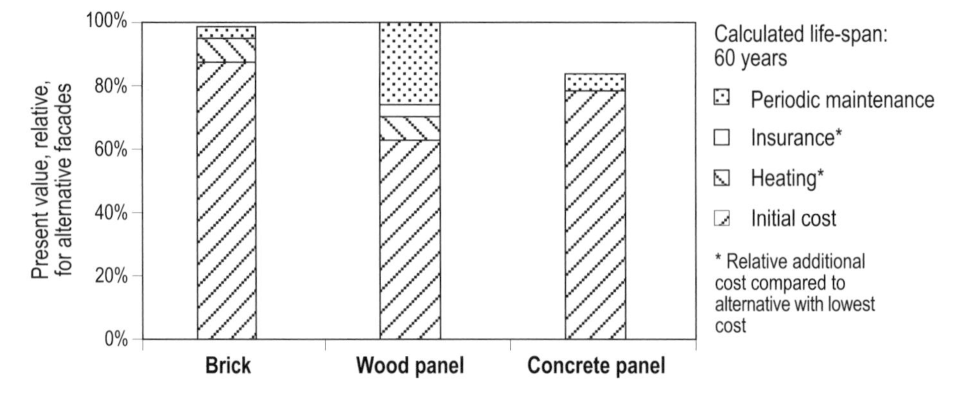

Fig. 3.3 Life cycle cost comparisons of three alternative façade designs.

Implementation

In relation to the potential advantages provided by ILCD, a possible drawback regarding increased design cost due to added procedures should be negligible. LCC and LCA could be naturally integrated in the routine production cost calculation from the early programming phase through the subsequent design phase. There are interesting opportunities for adding maintenance plans and other aspects of facilities management that would reduce the information gap between designers and clients. Feedback from the operation of buildings could be conveyed back to the designers to refine the data they use for ILCD calculations. The development of integrated IT solutions will facilitate the process issues but there are also administrative aspects that must be considered such as the opportunity to collect and access life cycle data through a common database.

Conclusions

Life cycle aspects should be addressed to a greater extent in the construction sector. The impact of the built environment regarding, for instance, energy use, waste streams and economy justifies a serious effort in the design phase to optimise certain characteristics of a building over its life cycle.

The tools and data for LCC calculations and for predicting energy use are available. Data regarding environmental aspects for products, namely Environmental Product Declarations, are being prepared. Methods for multiple criteria decision-making and ideas about rationalisation and integration of design and production can be imported from manufacturing industry. Several cases, for example applications of *open building*, show that the potential savings regarding economy, as well as the environment, can be significant and that improvements in function can be obtained through the adoption of life cycle design.

One barrier that must be overcome is how to introduce life cycle design into the regular design process in a practical manner. It is anticipated that new integrated IT solutions can be expanded with life cycle design tools and that the data and systems could also, in a logical way, be utilised in the facilities management phase to bring much added value. Feedback of information to the design team should then be secured and the oft-debated link between designer, supplier, producer and user could be reinforced.

References

Adalberth, K. (2000) *Energy Use and Environmental Impact of New Residential Buildings*, Report TVBH-1012, Lund: Lund Institute of Technology, p. 44.

Akao, Y. (1990) *Quality Function Deployment, QFD, Integrating Customer Requirements into Project Design*, Cambridge, MA: Productivity Press, pp. 1–24.

Andersson, K. (1998) *Livscykelanalyser av avloppsör – en jämförande analys av fyra utförda studier*, Gothenburg: Chalmers Industriteknik (in Swedish).

Arthursson, A. & Sandesten, S. (1981) *Årskostnadsberäkning – metoder.* Byggnadsstyrelsens rapporter, 1981–12, Stockholm: Kungliga Byggnadsstyrelsen (in Swedish).

Ashworth, A. (1993) How life cycle costing could have improved existing costing. In: *Life Cycle Costing for Construction* (ed. Bull, J.W.), London: Chapman & Hall, pp. 119–33.

BK, Byggsektorns Kretsloppsråd (2001) *Byggsektorns betydande miljöaspekter, Miljöutredning för byggsektorn*, Final Report, Stockholm: Byggsektorns Kretsoppsråd (in Swedish).

BKD, Byggkostnadsdelegationen, Swedish Delegation on Building Costs (2000) *Från byggsekt till byggsektor*, SOU 2000:44 (in Swedish).

Dale, S.J. (1993) Introduction to life cycle costing. In: *Life Cycle Costing for Construction* (ed. Bull, J.W.), London: Chapman & Hall, pp. 1–22.

Flanagan, R. & Norman, G. (1983) *Life Cycle Costing*, London: Royal Institution of Chartered Surveyors.
Flanagan, R., Norman, G., Meadows, J. & Robinson, G. (1989) *Life Cycle Costing: Theory and Practice*, Oxford: BSP Professional Books.
Gargione, L.A. (1999) Using Quality Function Deployment in the Design Phase of an Apartment Construction Project, *Proceedings: IGLC-7*, Berkeley: University of California, pp. 167–77.
Gluch, P. (2000) *Managerial Environmental Accounting in Construction Projects*, Paper II, Gothenburg: Chalmers University of Technology, p. 2.
Griffin, J.J. (1993) Life cycle cost analysis: a decision aid. In: *Life Cycle Costing for Construction* (ed. Bull, J.W.), London: Chapman & Hall, pp. 135–46.
Habraken, J. (1972) *Supports: An Alternative to Mass Housing*. London: The Architectural Press.
Habraken, J. (1981) *The Grunsfeld Variations: A report on the Thematic Development of an Urban Tissue*, Cambridge, MA: MIT Laboratory for Architecture and Planning.
Johansson, C. & Öberg, M. (2001) Life Cycle Costs and Affordability Perspectives for Multi-dwelling Buildings in Sweden, *Proceedings: 2nd Nordic Conference on Construction Economics and Organization*, Gothenburg: Chalmers University of Technology, pp. 287–97.
Kendall, S. (2000) *Residential Open Building*, London: E. & F.N. Spon.
Kirk, S.J. & Dell'Isola, A.J. (1995) *Life Cycle Costing for Design Professionals*, New York: McGraw-Hill.
Nieminen, J. & Houvila, P. (2000) QFD in Design Process Decision-Making. *Proceedings: RILEM/CIB/ISO Symposium ILCDES 2000* (ed. Sarja, A.). Helsinki: RILEM, pp. 51–62.
RICS (1980, 1986) *Pre-contract Cost Control and Cost Planning, A Guide to Life Cycle Costing for Construction*, London: The Royal Institution of Chartered Surveyors.
Rutter, D.K. & Wyatt, D.P. (2000) *Whole Life Costing: An Integrated Approach,* Proceedings: RILEM/CIB/ISO Symposium ILCDES 2000, Helsinki, pp. 113–16.
Saari, A. (2001) A systematic control procedure for environmental burdens of building construction projects, *Proceedings: 2nd Nordic Conference on Construction Economics and Organization*, Gothenburg: Chalmers University of Technology, pp. 107–15.
SABO, The Swedish Association of Municipal Housing Companies (1999) *Underhållsnorm 1999*, Stockholm: SABO (in Swedish).
SAR 65 (1965) *Basic method for designing residential supports without predetermining the size or layout of dwellings*, Eindhoven: The Foundation for Architects' Research, SAR.
SAR 73 (1973) *A methodology for the design of urban tissue*, Eindhoven: The Foundation for Architects' Research, SAR.
Sarja, A. (1998) Integrated life cycle design of materials and structures, *Proceedings: CIB World Building Congress 1998*, Gävle, pp. 827–34.
Sarja, A. (2000) Integrated life cycle design as a key tool for sustainable construction, *Proceedings: RILEM/CIB/ISO Symposium ILCDES 2000* (ed. Sarja, A.), Helsinki: RILEM, Keynote lecture, pp. 1–5.
SCB, Statistics Sweden (2000) *Yearbook of Housing and Building Statistics 2000*, Stockholm: Statistiska Centralbyrån.

SETAC (1993) *Guidelines for Life Cycle Assessment: A Code of Practice*, Belgium: SETAC-Europe.
Sjöström, C. (2000) Challenges of sustainable construction in the 21st century, *Proceedings: RILEM/CIB/ISO Symposium ILCDES 2000* (ed. Sarja, A.), Helsinki: RILEM, Keynote lecture.
Sterner, E. (2000) *Life-cycle costing and its use in the Swedish building sector.* Building Research & Information, **28**, 5/6, pp. 385–91.
Vägverket, Swedish Public Roads Authorities (1999) *Samhällsekonomiska kalkylvärden*, Planomgång 2002–2011, Report 1999: 170, Borlänge: Vägverket (in Swedish).
Yashiro, T. (2000) Development of open building system as sustainable urban element, *Proceedings: RILEM/CIB/ISO Symposium ILCDES 2000* (ed. Sarja, A.), Helsinki: RILEM, pp. 260–65.

Chapter 4

Life Cycle Costs of Commercial Buildings – a Case Study

Eva Sterner

Introduction

The environmental impact of buildings should be considered from a life cycle perspective since impact arises at different points; for example, when raw material is extracted, when materials, components and products are manufactured, when the products are used and then handled after use, and during transportation in all these stages. Different environmental loads are involved, but operational energy use is considered to be the largest single source of environmental impact from a building (Ecocycle Council 2000). The European Parliament has published a draft directive on the energy performance of buildings (Council of the European Union 2001). The purpose of the directive is to promote improvement in the energy performance of buildings, taking into account climatic and local conditions as well as indoor climate requirements and cost effectiveness. The directive concerns, among other things, energy certification of buildings and is likely to increase further the importance of energy reductions.

To highlight the relevance of energy use in buildings, Adalberth (2000) examined the life cycle energy for seven new residential buildings in Sweden and found that approximately 15% of total energy use was embodied energy and 85% was related to the operational phase. Similar conclusions are made by Cole & Kernan (1996), who examined the energy use for typical office buildings in Canada with a projected life of 50 years and found that 80–90% of total energy use was related to operational energy. When operational energy is reduced, for example through efficiency improvements, embodied energy becomes the dominant factor. Conclusions from a study in Australia of an environmentally-designed, commercial office building show that embodied energy is significant relative to operational energy use (Fay *et al.* 2000; Treloar *et al.* 2001). Clearly, both operational energy use and embodied energy should be reduced.

Reduced energy use has an effect on costs while other aspects of environmental building design can be difficult to promote without showing that costs are not substantially increasing. Barlett & Howard (2000) undertook a survey among quantity surveyors in the UK, and found that energy-efficient and environmentally-designed buildings were believed to increase initial cost by 5–15%. According to Barlett & Howard, however, the additional costs for an environmentally-designed building should not have to be more than 1%. The misconception over cost can stem from comparing high-profile buildings with conventional norms for building costs. Furthermore, a direct economic comparison of environmentally-designed and conventional buildings is difficult to perform as the environmental buildings often have a higher standard, using more exclusive materials, and lower maintenance costs. Basing cost appraisals on a life cycle perspective, rather than focusing on the initial investment cost, can be a way of establishing a better overview of all the costs and lead to more environmentally aware design.

This chapter reports on the life cycle cost of three new commercial buildings with high environmental targets, presented as a major case study. In particular, the effects of initial cost increases in relation to operational cost savings are addressed and compared with conventional buildings. The objectives of the study are to examine how the life cycle costs for the case study buildings differ from equivalent, conventional buildings where no explicit environmental consideration is taken.

Methodological considerations

Performing a life cycle cost analysis of a building requires the collection of a variety of information. Initially, a visual inspection of the three case study buildings was undertaken, followed by a meeting with the architect, to comprehend the planning and design process adopted and the approach to the environmental features. Further meetings with the contractor and the consultants for the services installations followed in order to obtain information about the construction work, the heating and ventilation system, and their associated initial costs.

In order to establish the maintenance cost and other annual costs, such as cleaning etc., all interior and exterior surfaces were calculated from original drawings. Future provision for maintenance was estimated according to intervals given in REPAB (2000), that also provides recommended values for annual costs. The energy use for the buildings was obtained from the consultants, based on the first year of operation and verified through calculations using the energy analysis program *ENORM* (Equa 2001), which is commonly used by authorities, consultants and contractors in Sweden.

Calculation of the present value of future costs was performed using a standard computer spreadsheet tool. For comparison, information on initial and operational costs for buildings of an equivalent size, without explicit environmental consideration, was obtained from the investor and the tenants.

Case study description

The case study buildings are collectively referred to as Greenzone and were designed, by Anders Nyqvist Arkitektkontor AB, to include three commercial facilities. The buildings are located on the Swedish east coast, in the city of Umeå, and were completed in the spring of 2000. Umeå is situated at a latitude of 64° and is subject to 4800 heating degree days annually (Abramson 1982).

The tenant of Building 1 offers various services related to automobiles. The building floor area is 3350 m^2 and includes offices, a car workshop, storage and sales areas split between two floors. Figure 4.1 shows the south-facing façade with an 80m^2 sun-catcher heating the incoming ventilation air. The east façade consists of large window areas to reduce the need for electric lighting. On all three buildings, a green roof, i.e. a roof with a layer of plants, is used to hold rainwater and to cool the buildings in the summer. For the purpose of comparison, the tenant provided data for a conventional building (Reference building 1) of an equivalent size located in Örnsköldsvik, 150 km south of Umeå and also on the east coast.

The tenant of Building 2 (Fig. 4.2) is an oil company that provides various services for motorists. The building's floor area is 590 m^2 and includes a sales area, a grocery store, a car wash and a petrol station. The energy demand for freezers, refrigerators and the car wash is greater than the energy requirement for heating and ventilation. For the comparative study, the company provided data on a conventional building of an equivalent size and floor plan located in Hudiksvall, 350 km south of Umeå (Reference building 2).

Fig. 4.1 South-facing façade of Greenzone Building 1.

Fig. 4.2 Greenzone Building 2.

Building 3 (Fig. 4.3), which includes a fast-food restaurant with kitchen and staff areas, has a floor area of 310 m^2. The energy demand in the building is largely represented by electricity for ventilation, grills, deep fryers, refrigerators and freezers. For the comparative study, data were obtained for a conventional building of an equivalent size and floor plan located in Härnösand, 250 km south of Umeå (Reference building 3).

All buildings are designed to reduce environmental impact, when compared with conventional buildings. Some general examples of environmental strategies are reusable building material, minimisation of construction waste and techniques for saving energy, water and electricity. The structure of the buildings has been designed to enable future reuse of material by making them easy to dismantle. A section of the wall and roof structure is shown in Fig. 4.4, where materials were selected for their environmental design qualities.

Other aspects that have a bearing on the environmental design of the buildings are, for example, windows made from heartwood frames, which is supposed to extend their life; and daylight, which is allowed to pass through these large window areas to reduce the need for artificial lighting. Daylight is also introduced through lantern skylights that are built into the roof. In

Fig. 4.3 Greenzone Building 3.

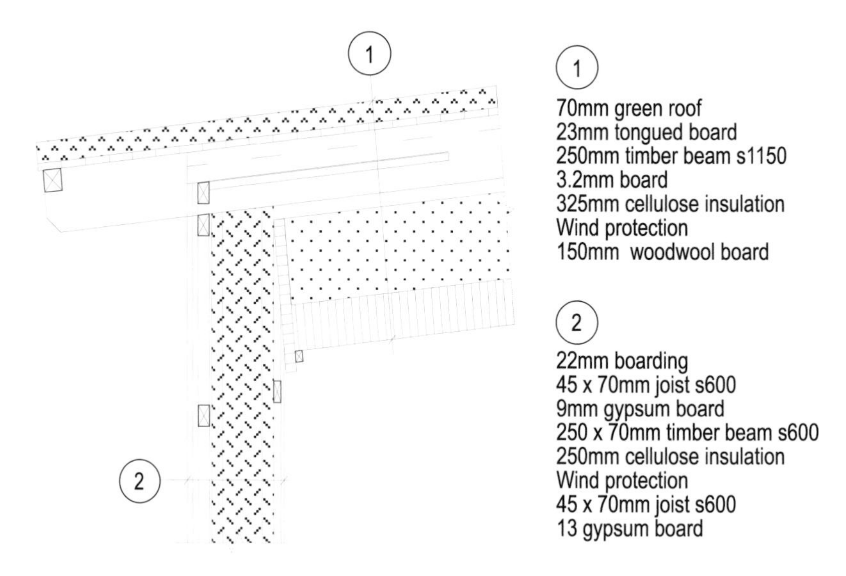

Fig. 4.4 Section of wall and roof of Greenzone buildings.

order to reduce the need for water, extensive work has been carried out in the surrounding area where creeks have been created to collect and purify surface water. The groundwater level at the site is high, making it possible to use groundwater for irrigating the area as well as supplying low water-consuming vacuum toilets and the car wash.

Energy conservation

Several measures have been taken to reduce energy demand in the buildings. One important aspect is the recovery of heat; another is the use of different sources of energy, including ground-source technology, recycled surplus heat, solar energy and heat pump technology.

Heating and cooling is delivered partly by a ground system of five holes drilled down to 150 m below the surface and connected to a heat pump. The heat pump has an efficiency factor of 3.3, i.e. for each kWh of electricity used for operating the heat pump 3.3 kWh of heat is produced. The heat pump is connected in series to an electrical boiler that is used to add additional heat during peak periods. The heat pump is, however, expected to cover 95% of the energy demand for heating.

Culverts connect the buildings and are used to recycle surplus energy. This is collected from heated coolant water from the refrigerator system in Building 2 and from grills and deep fryers in Building 3. A sun-catcher is installed on the south-facing façade of Building 1, heating the incoming ventilation air during the period September to May.

In Building 1, heat is distributed by an under-floor heating system; in Building 2, a small area of under-floor heating is used, together with radiators; and in Building 3, heat is supplied solely by radiators since under-floor heating is ineffective when the window area is large.

In order to determine the reduction of energy needed for heating the new buildings, for comparison with conventional buildings, calculations were provided by the HVAC consultant – see Table 4.1.

For Greenzone, the total energy demand – indicated by purchased electricity – has been measured over a full year from 1 April 2000 to 31 March 2001. For the conventional buildings, data on annual electricity consumption was obtained. For the Hudiksvall and Härnösand buildings, heat is provided by electricity to convector air heaters. In the Örnsköldsvik building, district heating and radiators are used. Table 4.1 shows the energy performance of the buildings.

Table 4.1 Energy demand for Greenzone and the conventional buildings.

	Case 1		Case 2		Case 3	
Operational energy	GZ 1	Ref 1	GZ 2	Ref 2	GZ 3	Ref 3
Floor area (m²)	3350	4800	590	590	310	310
Energy for heating						
District heating (kWh)	0	675000	0	0	0	0
Electric heat pump and boiler (kWh)	53596	0	62010	0	34222	0
(kWh/m²)	16	141	105	0	110	0
Electricity						
(kWh)	210000	500000	432290	600000	318878	450000
(kWh/m²)	63	104	733	1017	1029	1452
Total energy demand						
Electricity (kWh/m²)	79	104	838	1017	1139	1452
District heating (kWh/m²)		141				

Costs are shown in SEK (€1 = SEK 9.02)

The electricity needed for the daily operation of the buildings is assumed to have little or no dependence upon the actual design of the buildings, with the exception of electricity for lighting and ventilation. Lighting need is reduced for all Greenzone buildings due to the lantern skylights and motion sensors.

The energy demand for the buildings is based on measured values representing a normal year. Buildings 2 and 3 are electricity demanding, but since waste heat is produced only a small amount of energy is needed for heating. The energy for heating has, therefore, not been separated for Reference buildings 2 and 3 in Table 4.1, but is included under common electricity for these.

Life cycle cost analyses

The life cycle costs of an asset can be defined as the total cost of that asset over its operating life, including initial acquisition cost and subsequent running costs. The major life cycle cost components are, according to Flanagan & Norman (1983):

- capital cost, I, i.e. initial cost, including site costs, design fees and building cost;
- operating cost, O, i.e. annual costs including cleaning and energy;
- maintenance cost, M, i.e. annual costs and costs for replacement and alteration; and
- salvage value, S.

Of these costs, it is only the initial cost that is calculated as a present value, while the other costs are first calculated then discounted to present values to provide a common base in time. The total cost in equation (4.1) below is calculated as the sum of the initial cost and the present value of operating and maintenance costs for the selected period of analysis (Bejrum 1991).

$$\mathrm{LCC} = I_0 - S_N + \sum_{n=t}^{N} (O_t + M_t) \cdot \frac{1}{(1+r)^t} \tag{4.1}$$

The life cycle cost evaluation performed here is defined to include initial costs, maintenance and operational costs. The life span, N, is assumed to be 50 years since the economical life is normally set to 40–50 years. Also, Swedish building codes for structures use 50 years as the typical life of a building. Many buildings have, however, a significantly longer technically useful life. The salvage value, S, represents the net sum to be realised from disposal of an asset at the end of its economic life, at the end of the study period or whenever it is no longer to be used (Ruegg 1978). If salvage value occurs at the end of a long study period, it tends to have relatively little weight in the analysis, due to the diminishing effect of the discounting operation. Therefore, salvage is not included in the evaluation. Finally the discount rate, *r*, has been set to 4%, which has been selected to correspond to the long-term cost of borrowing money. The discount rate is a major factor of uncertainty and can be varied in a sensitivity analysis.

Initial cost

Initial cost is the investor's cost that arises directly from the project, including the cost of land, acquisition fees, design team fees, demolition and site clearance, construction costs and furnishings. For Greenzone, initial construction costs are calculated from invoices supplied by the general contractor. Design costs and costs for services installations related to heating, electricity etc. were provided by the HVAC consultant as a lump sum, and have been divided between the buildings in proportion to their construction cost. Figure 4.5 shows the distribution of initial costs.

The dominant elements of the initial costs for the Greenzone buildings are building cost, services installations cost, and land and site costs, representing 47%, 27% and 17% respectively. The building costs for the three reference buildings were compared with those in Greenzone. A cost increase of 5–17% has been found. These increases can be traced to higher services installation costs due to: interaction of the heating and cooling system, making it possible to use surplus energy; installation of vacuum toilets; façades of screwed timber panels (instead of nailed panels); wood fibre insulation in exterior walls and roofs; a layer of plants on the roof; lantern skylights; plants to purify the indoor air; and the sun-catcher. Also, the land and site cost is higher than for conventional buildings, because of extensive work in the area surrounding the facilities.

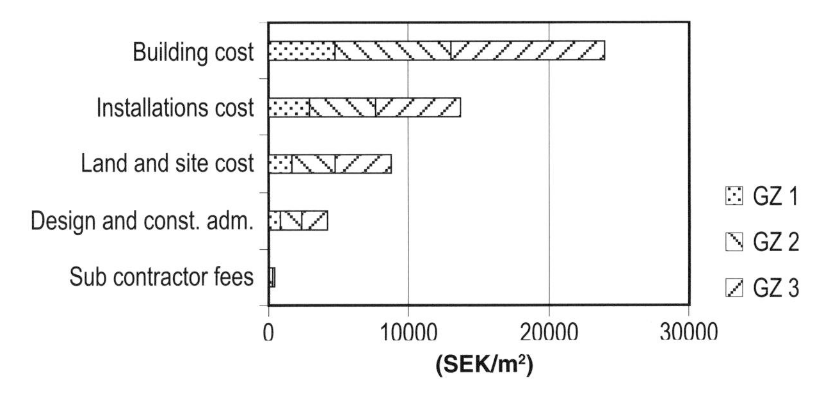

Fig. 4.5 Distribution of initial costs for Greenzone buildings.

Operational cost

The operational cost of a typical commercial facility is mostly represented by the variable costs of energy and cleaning and, to a lesser extent, by fixed costs such as rates, insurance, security, and management or administration. Maintenance of the grounds, buildings and services installations, which is performed annually, forms part of the caretaking or facilities management function which, along with refuse disposal, administration and insurance, is assumed to have little or no dependence upon the design of the buildings. These costs are deemed to be the same for both the Greenzone and conventional buildings. The data required for calculating these has, for the most part, been estimated here by using published statistical data from REPAB (2000). As shown in Fig. 4.6, the energy costs representing the total amount of electricity purchased, followed by cleaning, are the most significant of the annual costs. Of the energy cost, approximately 25% is used for heating and ventilating the buildings. For Building 3, the costs of water and refuse disposal are relatively large because of the nature of the activities in the building.

Maintenance cost

Maintenance needs should be kept under constant review to ensure that the building is in an acceptable state of repair. The cost of repairs and replacement can vary widely, depending on the state of the building and how users look after it. In the early days of use, costs are less than when the building has begun to deteriorate (Flanagan & Norman 1983). The costs for maintenance have been classified into:

- envelope (external walls, roof, windows and doors);
- internal finishes (walls, floors, ceilings, windows and doors); and
- mechanical and electrical system (heat pump, boiler and ventilation).

On the assumption that no substantial reconstruction is undertaken, the maintenance of the main structure can be omitted since the structure's life is the same as the analysis. Also, the effects of modernisation and adaptation of the buildings are not included. For remaining elements, the frequency and cost of replacement was found in REPAB (2000). The present values of the maintenance costs are shown in Fig. 4.7.

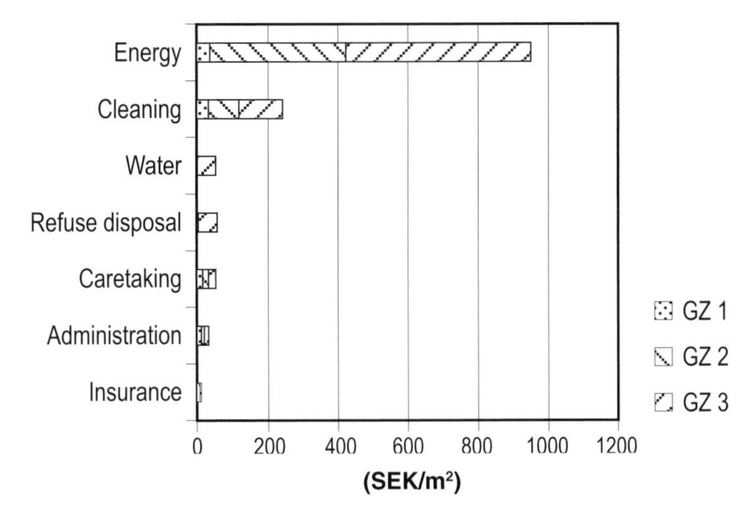

Fig. 4.6 Annual operating costs for the Greenzone buildings.

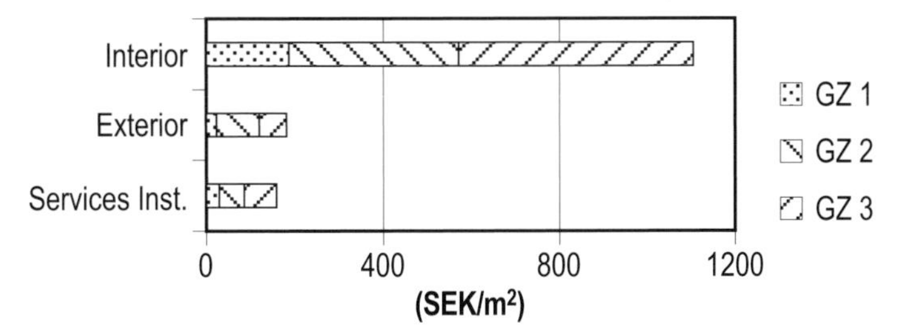

Fig. 4.7 Present values of maintenance costs for Greenzone buildings at a 4% discount rate.

Whole life cycle cost

To determine the building's whole life cycle cost, a period of analysis of 50 years was used. A discount rate of 4% was selected as this corresponds to the long-term cost of borrowing money. Initial costs were calculated as a lump sum, present value occurring at the beginning of the base year. Operational costs are evaluated as lump sum amounts at the end of each year and discounted to present values. One other simplification is that maintenance costs are assumed to be equal for Greenzone and the conventional buildings, even though some differences do exist. It is also assumed that the price attached to operations and maintenance will change at about the same rate as prices in general, i.e. they will remain the same over the study period. Costs for maintenance are treated as a lump sum at the end of the year in which they are estimated to occur. The life cycle cost and each of the included costs are presented in Table 4.2.

Table 4.2 Comparative present values of costs in SEK/m².

	Greenzone			Conventional buildings		
Cost	GZ 1	GZ 2	GZ 3	Ref 1	Ref 2	Ref 3
Initial	10053	17592	22903	8355	16180	20000
Energy (PV)	784	8351	11353	2612	10137	14469
Other operations (PV)	1486	2645	5823	1486	3810	5823
Maintenance (PV)	239	536	678	239	536	678
Life cycle cost	12562	29124	40757	12692	30663	40970

Table 4.2 shows that the present value of the whole life cycle cost for the environmentally-designed buildings is approximately in the same range as for the conventional buildings. From the life cycle cost assessment perspective, the environmentally-designed buildings represent a better proposition than the conventional buildings. The energy used for heating and ventilation of Reference building 1 has been calculated and a reduction of costs of 60% has been found. If looking from a whole energy use perspective, the reduction of costs for heating and ventilation is of minor significance to the more energy-demanding tenants, although they are significantly reduced. The initial cost is dominant over the operating and maintenance cost for the environmentally-designed buildings, but for the conventional buildings operating and maintenance costs are close to initial costs.

Conclusions

The results of the study show that the environmentally-designed buildings are in the same cost range as the conventional buildings, if adopting a life cycle cost perspective. This shows the importance of using a longer investment horizon in order to justify higher initial costs, which for the conventional buildings are 5–17% lower. Costs for products with environmental design qualities can be higher, but with increasing demand and better competition between manufacturers this market distortion can be corrected. The operating costs for Greenzone are lower than for the traditional buildings examined in the study. This has a significant effect on the total life cycle cost. To a large extent, energy reduction is in the hands of tenants, although real incentives for reducing consumption are not yet present. For environmentally-designed buildings, the operational costs, especially related to energy, are significantly lower than the initial investment costs at somewhere in the region of 58–84% of the total life cycle cost.

References

Abramson, B. (1982) *Handboken Bygg, Husbyggnader och Installationer,* Stockholm: LiberFörlag (in Swedish).

Adalberth, K. (2000) *Energy use and environmental impact of new residential buildings,* Doctoral thesis, Lund: Lund Institute of Technology.

Barlett, E. & Howard, N. (2000) Informing the decision makers on the cost and value of green building, *Building Research & Information,* **28**, *5/6,* pp. 315–24.

Bejrum, H. (1991) *Livscykelekonomiska kalkyler för byggnader och fastigheter,* TRITA-FAE 1033, Stockholm: Royal Institute of Technology (in Swedish).

Cole, R. & Kernan, P. (1996) Life-cycle energy use in office buildings, *Building and Environment,* **31**, *4,* pp. 307–16.

Council of the European Union (2001) Proposal for a Directive of the European Parliament and of the Council on the energy performance of buildings, ENV 636, Brussels: Council of the European Union, 5 December.

Ecocycle Council (2000) *Byggsektorns betydande miljöaspekter,* Stockholm: Ecocycle Council (in Swedish).

Equa (2001) *ENORM version 1000,* Equa Simulation Technology Group.

Fay, R., Treloar, G. & Usha, I.-R. (2000) Life-cycle energy analysis of buildings: a case study, *Building Research & Information,* **28**, *1,* pp. 31–41.

Flanagan, R. & Norman, G. (1983) *Life Cycle Costing for Construction,* Oxford: BSP Professional Books.

REPAB (2000) *Årskostnader 2000: nyckeltal för kostnader och förbrukningar,* Stockholm: REPAB AB (in Swedish).

Ruegg, R.T. (1978) *Life cycle costing. A guide for selecting energy conservation projects for public buildings,* NBS Building Science Series 113, Washington DC: US Department of Commerce.

Treloar, G., Love, P. & Holt, G. (2001) Using national input-output data for embodied energy analysis of individual residential buildings, *Construction Management and Economics,* **19**, pp. 49–61.

Chapter 5

A Life Cycle Cost Approach to Optimising Indoor Climate Systems

Dennis Johansson & Anders Svensson

Introduction

An indoor climate system must fulfil some important functions. It must meet its design criteria, which means correct temperatures and clean air as required. Additionally, the system must consume the least amount of resources as possible from both an economic and environmental perspective. As a general rule, the construction sector in Sweden consumes about 40%, or about 154 TWh, of the total energy used in the country (Energimyndigheten 2000). A major part of this is used to provide buildings with the necessary energy for heating and cooling. Another factor is that the design of indoor climate systems has been overly focused on initial or capital costs and less on operating costs.

With a life cycle approach, both the environmental and economic performance of the indoor climate system could be improved. Today, there is a lack of knowledge and tools for determining life cycle costs. Research is therefore needed. This chapter discusses how to create a model for life cycle cost (LCC) analyses of different indoor climate solutions and discusses the need for such a model.

State-of-the-art review

In order to create the right indoor climate system, it is important to decide on what people demand from their indoor climate. Ideas for establishing methods and devising tools for LCC analyses are also important. Here, the ongoing research as well as the existing tools will be discussed. In order to optimise the costs from the client's perspective, a life cycle approach is needed to deal with whole buildings or even several buildings and their locations, the tenants involved, social systems, migration and so on. To solve such a problem requires that life cycle analyses must be subdivided into smaller, more manageable calculations. It is important to decrease the complexity, but still keep the *big picture* in view.

First, research dealing with ventilation is reviewed. Second, life cycle costs for the building, as well as indoor climate systems, are discussed. The last section deals with examples and the impact of ventilation on work performance and health.

Ventilation principles and functionality

The basic types of ventilation systems available today include natural ventilation, mechanical exhaust ventilation and mechanical supply-exhaust ventilation. The latter can incorporate a heat exchanger. Natural ventilation uses thermal driving forces and wind pressure. It works best in the wintertime, is very quiet and needs little maintenance. Problems with this system can include weather dependency and the difficulty of controlling ventilation rates. In some cases, it is impossible to achieve sufficient airflow rates. Natural ventilation is less common in today's design, but mixed systems known as hybrid ventilation are increasing and a number of research projects are investigating their performance.

Mechanical exhaust ventilation draws air from outlet devices that are located in the kitchen and the bathroom where the air is most polluted. This creates a lower pressure inside the building, with supply air coming through inlet devices in the wall. A common problem is that an open window, for example in the kitchen, changes the pressure balance resulting in zero airflow in a bedroom when the bedroom door is closed (Engdahl 1999). Exhaust ventilation can be fitted with a heat recovery unit, providing a positive influence on the system's life cycle cost.

Exhaust and supply ventilation has two fans supporting both exhaust and supply air. Here, it is easy to add a heat exchanger for heat recovery from the outlet air. Normally, the exhaust airflow rate is slightly greater than the supply rate so that moisture is not driven through the walls. A problem is that the exhaust filter must be cleaned or replaced more often than the supply filter, because of the more polluted indoor air; if it is not, the exhaust airflow rate will decrease. Exhaust and supply air systems are a more expensive solution, but provide the best control over flow patterns and flow rates (Engdahl 2000).

Functionality is an essential feature of an indoor climate system and the main reason for implementing it. In Sweden, as in other countries, there are recommendations and legal demands setting the airflow rates and temperatures in buildings. Mandatory Swedish ventilation control regulations reflect experiences from practice. They imply that research about the ventilation process must focus on questions regarding the human need for good indoor air quality as well as for a good indoor climate without noise and draughts.

More than half of all schools, offices and multi-dwelling buildings failed their mandatory ventilation control test according to Engdahl (1998): mandatory ventilation controls started in 1992 and the client must attend to any faults. The worst cases were natural ventilation systems.

It is important that a ventilation system continues to operate for its design life. The system must be self-adjusting, or it must be adjusted whenever a change is made, or age alters its performance. Office buildings are prone to frequent refurbishment and hence a lot of money can be spent on altering the ventilation system. However, a flexible system can be self-adjusting and can accommodate new demands brought about by refurbishment.

The literature abounds with papers on room air distribution, numerical methods and indoor air quality. For example, the energy efficiency of displacement and mixing ventilation has been compared in an environmental chamber study, and displacement ventilation was shown to be the best (Awbi 1998). Published studies can be helpful when deciding if a proposed ventilation system is likely to work properly in a given situation. If it is unlikely to work, there is no need to calculate the LCC.

A related project to the one discussed in this chapter is ongoing. This deals with determining optimal control strategies to achieve lower energy use and better functionality. For example, by controlling the flow rate and the supply air temperature for an exhaust and supply air system, the energy use, and thereby the LCC, can be decreased. Today, control strategies seem to be simplified as much as possible. One major result so far is that condensation occurring by air cooling is an important factor.

LCC principles, tools, use and research

Whole life cycle cost, or simply life cycle cost (LCC), is the total cost a client pays for a product. The LCC consists of the investment cost, energy cost, maintenance cost, demolition cost and all other recurrent costs, which are normally discounted to today's value, known as the present value, by the use of an estimated interest rate.

Difficulties with LCC analyses include uncertain input data that depend upon human actions within society as well as variability in technical performance. The lifetime of the product is not easy to decide upon and much depends on processes such as changes in business activity and technical limitations. Interest rates are extremely difficult to predict with any certainty. Energy cost, outdoor climate and client behaviour are other uncertain aspects (Bull 1993; Flanagan *et al.* 1989). *How many different costs should be taken into account?* For example, improved work performance due to better indoor air quality can be significant, but is seldom taken into account.

Life cycle thinking, based on experience as opposed to formalised techniques, is often used though not to its full potential. For example, disposable items are not used in a kitchen because they would be more expensive in the long run. A life cycle approach has also been used in other sectors like national defence, road construction and agriculture. Often, in the construction sector, some sort of life cycle perspective is taken, but this tends to rely on old experience (or even perceptions) of good solutions. This may not be the best approach. In the construction process, the developer meets the investment cost, but does not guarantee long-term functionality. There are no incentives for the developer to undertake LCC analyses, so long as clients do not require them.

To give an example, one survey has indicated that 40% of municipal construction projects in the US were based on LCC analyses. The reasons given for not using LCC analyses were lack of guidelines and the difficulty of estimating future costs (Arditi 1996). A survey undertaken in 1999 shows that 66% of clients surveyed use life cycle cost analyses to some extent in building projects (Sterner 2000). Life cycle cost analysis is rarely adopted in all phases of the construction process. Lack of tools and knowledge is cited as a significant reason.

Interest in making LCC analysis routine has resulted in many practical guidelines and forms. One such example is *Energy Efficient Procurement, ENEU 94* (Sveriges verkstadsindustrier 1996), which is a tool for determining the life cycle cost of a fan system. *ENEU 94* takes much more than air handling into account; for example, electrical motors and machines. *ENEU 94* was updated in 2001, and is available on the web only. The idea is to calculate the LCC for given scenarios and to compare them. *ENEU 94* is also a form of guidelines with the aim of forcing the design consultants to undertake the analysis in order to secure the contract. The guidelines have been developed mainly to help determine the optimal size of an air handling unit: special control strategies for air handling plant have not been included. *ENEU 94* does not incorporate spatial costs or system functions. If a number of competing consultants make one or even several calculations each, a good solution might be found, but it is not guaranteed.

An EC-supported project is developing guidelines similar to *ENEU 94* and carries the name *LCC-based Guidelines on Procurement of Energy Intensive Equipment in Industries* (Eurovent 2001). In *ENEU 94* there is a client focus, but the EC project has a producer focus.

The Norwegian University of Science and Technology (Vik 2001) has an ongoing project dealing with life cycle costs for natural ventilation systems. One aim of this project is to determine when natural and hybrid ventilation results in a low life cycle cost.

LCC analyses for indoor climate systems

LCC analyses are increasingly undertaken by researchers, consultants and clients to determine the best ventilation system. These analyses most often compare two or more types of systems, but they do not come up with a completely optimised solution. Of the energy used over the lifetime of a building 85% is consumed during the occupation phase. The majority of this is due to the building's heating and cooling systems (Adalberth 2000). The energy used within an indoor ventilation system is the balance between energy saving and sufficient airflow rates for a clean indoor environment.

Office cooling can be achieved either by an all-air system or radiant cooling system. It is generally expected that a radiant cooling system uses less energy than an air system because of decreased air movement. A saving potential of 10% was found in a Japanese study (Imanari *et al.* 1999). In another example, a chilled ceiling was used in an office building, providing a slightly better indoor climate in addition to the energy saving. A German study (Sodec 1999) found a lower LCC for radiant cooling ceilings, when compared with an all-air system, assuming the systems are well designed and operated correctly.

In the US, 30% less energy use for a radiant system was found. Peak power use was also discussed, a decrease in which could reach 27% (Stetiu 1999). The peak power cost can be taken into account in the LCC analysis. In the US, as in other countries with hot climates, a lot of cooling is required compared with Sweden and northern Europe, where power supply problems appear in the winter. Then, the heating peak power is of greater interest.

A deficiency revealed in these studies was that the all-air system could be improved by better control strategies. Also, the features with increased airflow from the all-air system were not discussed; neither was heating discussed and this is an important part of energy use in northern Europe. There remains a need for a more systematic approach to designing complex indoor climate systems.

A comparison of radiant cooling and all-air cooling has been discussed in the context of a life cycle assessment approach (Johnsson 2001). In a life cycle assessment environmental impact, obtained from the assumed danger of sub-activities, determines the system instead of its costs. Johnsson tested two different methods to provide comparative results.

There has been some work dealing with the optimisation of duct sizes (Besant *et al.* 2000; Jensen 2000). Efforts were made to describe both the energy costs and the materials and installation costs, to yield an optimised pipe system. This is usable, but whole indoor climate systems or combinations of systems were not optimised.

A central cooling system was compared with a split system in a 29-storey office building in Hong Kong. The central system gave the lowest LCC (Yang *et al.* 2001). However, Yang *et al.* only discuss the comparison between two scenarios; there could be an even better solution, so the general system approach is still missing.

Matsson (1999) has made some life cycle cost calculations for systems with different control strategies. Three systems were compared:

- with a constant air volume flow (CAV);
- with a variable air volume (VAV) with constant pressure in the main duct; and
- with a variable air volume with changeable pressure in the main duct in order to minimise the static pressure.

The analyses were based on the real costs for a school and an office building. The system with changeable pressure in the main duct shows the lowest life cycle cost in both cases due to the lower pressure drop decreasing the energy used.

The heating or cooling system can also include the production of power. The heating system for housing was found to be of great importance when determining insulation thickness (Gustafsson 2000).

Work performance and health aspects

Increased performance caused by better thermal climate has been investigated and a very small increase in work performance is easily profitable. This clearly makes air conditioning in offices a good investment. The same argument is valid if, for example, higher airflow rates result in better health or increased work performance.

Of the research performed in the area of thermal climate, little has actually been completed, but some is ongoing. For instance, it has been shown that a significant increase in work performance occurs when increasing the airflow rate (Wargocki *et al.* 2000). Reactions between ozone and volatile organic compounds are suspected to result in more dangerous substances. With higher airflow rates, the reactions seem to be less noticeable (Weschler & Shields 2000). A third of several articles, in the same vein, show that a high level of absenteeism – attributed to sickness – is correlated to low airflow rate (Milton *et al.* 2000).

There is increasing evidence of a positive effect with higher flow rates and, perhaps, the legal requirements are too low. This aspect can be important when choosing between cooling with higher airflow rates compared with chilled ceilings. The claimed higher work performance and better health can make a great impact on life cycle costs.

Research project

Project description and objectives

In order to be able to investigate the lifetime cost for different technical indoor climate solutions, a tool and a model are needed to support any calculations. The aim of the project discussed in this section is to develop a calculation model for life cycle cost (LCC) analyses of different technical systems and solutions for ventilation and climate systems, taking into account flexibility and control strategies. The model will also be used on practical examples and be implemented in the sector.

Research methodology

The nature of the research within this project is both *positivistic* and *hermeneutic*: most problems with LCC analysis can be attributed to human factors and are in that way hermeneutic. Furthermore, there is no evidence that the past is able to describe the future. One method could be to try to determine dependencies for the human-related variables like lifetime, interests, energy cost and climate. This is a very uncertain way and not within the boundaries of this project. Another method can be to assume variables based on believable facts and perform sensitivity analysis to find the break points when the results change.

Describing building costs is of a similar nature. The most exact way can be to describe every cost for each variable and manipulate it over time; but something more transparent is needed. A model describing building costs must be created and validated by current experience, yet must remain flexible enough to accommodate future changes. Maintenance cost is also difficult to determine with certainty but experience combined with the manufacturer's product information should produce valid and reliable results.

Energy use and technical function are more positivistic by their nature except, possibly, outdoor climate changes. Mathematical expressions are needed in the model. These expressions can be obtained either by model building, or by case studies and measurements from existing buildings. Measurements from existing buildings can take time to accumulate and there are many parameters that can confound the result. It would be very difficult to ascertain if the result is reproducible based on the parameters measured. In this research project, model building will be used and measurements will confirm and test the validity and reliability of the model. Hopefully, new or better system ideas will be easier to model.

A robust model for performing LCC calculations must be able to handle some important demands. It should be able to be used on any kind of building and with any kind of fan ventilation system. Combinations of water-based or air-based systems must be handled as well as different control strategies. Building costs and volumes will be considered. Natural ventilation systems will not be considered because of the different technical principle.

The energy characteristics of the building will not, however, be handled by this model. The risk of separating building physics from the model is a non-optimised connection between the building and the indoor climate system. This should be taken into consideration, but other models may prove more suitable and should be used to calculate the energy demands for the different alternatives. That means data about the building such as orientation, internal heat loads, indoor climate demands and outdoor climate from these external models will give the input to the LCC model.

The model will be used by the person who designs the climate and ventilation system in a building. Therefore, it is important that the method provides a model and tool that are easy to use by design consultants and clients.

The next step will be to introduce the model and tool to the sector, whose needs it will be important to canvass. A survey has already summarised the experience and market-tested the potential for the project and tool. A goal of this survey was to create opportunities for co-operation between the project and construction sector. The survey was undertaken to estimate the use of life cycle cost analyses among clients, both public and private, and amongst consultants working with indoor climate systems. A questionnaire was sent out to one municipal building department in each of the 21 counties in Sweden, and to 20 private clients and 20 technical consultants. All recipients were selected at random; however, no claim is made as to the nature of the sample other than it is drawn from a population sharing similar characteristics. The questions investigated the extent to which life cycle cost analyses were used both when designing the building, and when designing the air handling plant in the building. The reasons for using or not using LCC analyses and related questions were also included.

Based on the above discussion, some methods have been selected. First, a model for calculating LCC will be developed. A starting approach is to base the model on a comfort standard; for example, ISO 7730. Later, a value can be placed on work performance, health and the system's interface. To build the model, it is necessary to define the available systems and investigate them technically to describe energy use, lifetime and space use. The functionality and opportunities for using a particular system in a given case must also be stressed. Information must then be collected about the (tool) user's needs, product costs, building costs and other issues influencing the LCC. In the beginning, the model will focus on office buildings to gain experience of its use and to obtain quicker results.

After building the model, example calculations have to be performed and verified against a real example. Another way would be to examine cases, but it could be difficult to implement new ideas and systems. Co-operation with other researchers is also important for obtaining easier verification and for fine-tuning the method. When the model is finished, example calculations will be performed as part of a sensitivity analysis and for drafting user guidelines. Here, an issue could be *how to decide between two systems having the same LCC*.

Reseach results and industrial impact

The use of LCC calculations

As stated above under *research methodology*, 61 questionnaires were sent out to investigate the use of LCC calculations when designing the services installations of buildings. The response rate was 39%. Of the total number of respondents, 31% used life cycle cost analyses to some extent. The use of LCC analyses was 43% by consultants, 33% by private sector clients and 20% by municipalities. Those who used the analyses most of all did so in 20% of their projects, though this was more common on larger projects. It was not possible to correlate the use of analyses to outdoor climate or company size. The response rate was low, perhaps because of a lack of interest. This suggests that the use of LCC in the sector is lower still.

Despite the low use of LCC analyses, 88% of respondents said that energy use is a very important environmental issue, while 34% needed better computer tools to perform the calculations. Lack of knowledge was given by 65% of respondents and 69% believe that the number of LCC analyses will increase in the future. A number of interesting aspects have been displayed and will be useful in the research project.

Examples of basic LCC calculation

Some of the problems associated with modelling climate and ventilation systems that adopt a life cycle perspective are discussed in two examples. The first example is the life cycle cost for two different ventilation systems in an assumed office cell. The systems are a mechanical exhaust and supply air system; first without heat recovery unit, and second, with a heat recovery unit. The heat exchanger causes a pressure drop and therefore needs more electrical power.

A calculation showed that the heating cost saved because of the heat exchanger in an assumed example of an office cell could be less than the added electricity cost due to the pressure drop in the heat exchanger. At high internal heat loads the heat exchanger in this example is harmful. This example may not

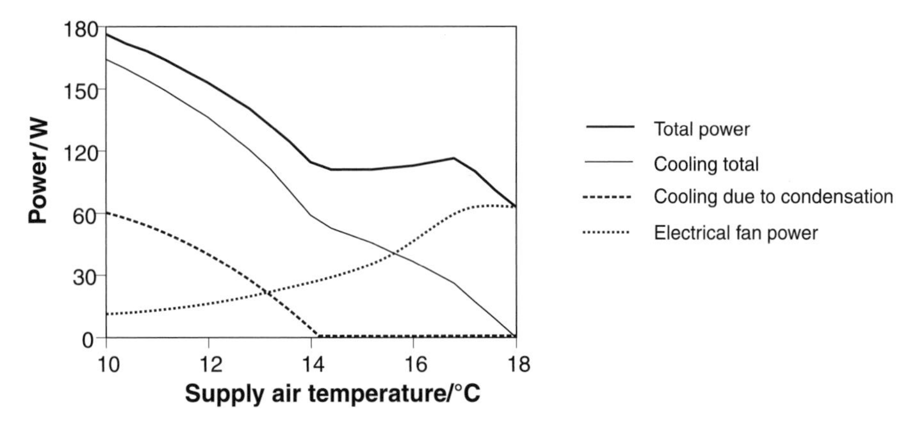

Fig. 5.1 Power needed to cool an office cell (depending on the supply air temperature).

be realistic. The best way could be to bypass the heat exchanger when it is not needed. Still, it is an example of the potential of more in-depth knowledge.

The second example concerns the cooling of a standard office cell with air. The power use as a function of the supply air temperature was examined. The internal heat load was assumed to be 500W, the specific fan power to be 2kW/(m^3/s), the room temperature to be 25°C, the heat transfer to be 20W/°C, the minimum airflow rate to be 10 l/s and the coefficient of performance to be 3. The outdoor temperature was 20°C and the relative humidity 75%. A low airflow rate needs a lower supply temperature with condensation as the result. Condensation power is non-linear. Figure 5.1 shows the result. The power minimum, providing the lowest energy cost in this condition, was found just before condensation occurred. The temperature was optimal in this case, but is seldom adopted in practice.

Conclusions

The results of the preliminary research are simple and do not represent reality in the manner of complexity and available solutions. Nonetheless, they show ideas and options that the project will handle. The use of life cycle cost analyses is low but increasing. Therefore it will be more important to know how to perform these calculations in the future.

The state-of-the-art indicates that there is a lack of implementation of the knowledge available today. A lot of climate systems do not work properly and are not as efficient as they could be. Additionally, clients and users do not have the competence or experience needed to determine if their indoor climate is good or not. A lot of money is spent on refurbishing inflexible systems, and sometimes they do not work efficiently anyway. This means that the research must secure the functionality of the indoor climate systems as well as providing the sector with good conditions for achieving them.

The influence on health and work performance by the indoor climate system will most likely be examined in the near future. The outcome of such research can be very difficult and expensive to implement in our indoor climate systems. Depending on the location, an increased airflow rate may result in higher work performance, a lower amount of sleep requirement at night or better health. If such effects can be proved then the systems need to be improved. It is important that the indoor climate systems built today handle the demands of tomorrow in an efficient and economical way.

References

Adalberth, K. (2000) *Energy use and environmental impact of new residential buildings*, TVBH-1012, Lund: Lund Institute of Technology.

Arditi, D.A. (1996) Life-cycle costing in municipal construction projects, *Journal of Infrastructure Systems*, **2**, *1*, pp. 5–14.

Awbi, H.B. (1998) Energy efficient room air distribution, *Renewable Energy*, **15**, pp. 293–99.

Besant, W.R., Eng, P. & Asiedu, Y. (2000) Sizing and balancing air duct systems, *ASHRAE Journal*, **42**, *12*, pp. 24.

Bull, J.W. (1993) *Life Cycle Costing for Construction*, London: Blackie Academic & Professional.

Energimyndigheten (2000) *Energiläget 1999*, ET 81:1999, Use and production of Swedish energy, http://www.stem.se [1 Sept. 2001] (in Swedish).

Engdahl, F. (1998) Evaluation of Swedish ventilation systems, *Building and Environment*, **33**, *4*, p. 197.

Engdahl, F. (1999) Stability of mechanical exhaust systems, *Indoor Air*, **9**, pp. 282–89.

Engdahl, F. (2000) Stability of Mechanical Exhaust and Supply Systems, *Roomvent 2000*, **2**, pp. 1207–12.

Eurovent (2001) Life cycle cost, *Review*, ed. Becirspahic, S., May 2001, s.becirspahic@eurovent-certification.com, [1 Sept. 2001].

Flanagan, R., Norman, G., Meadows, J. & Robinson, G. (1989) *Life Cycle Costing – Theory and Practice*, Oxford: BSP Professional Books.

Gustafsson, S.I. (2000) Optimisation of insulation measures on existing buildings, *Energy and Buildings*, **33**, pp. 49–55.

Imanari, T., Omori, T. & Bogaki, K. (1999) Thermal comfort and energy consumption of the radiant ceiling panel system. Comparison with the conventional all-air system, *Energy and Buildings*, **30**, pp. 167–75.

Jensen, L. (2000) *Injustering av ventilationssystem – underlag för implementering*, TABK – 00/7058, Lund: Lund Institute of Technology (in Swedish).

Johnsson, E (2001) *Projekteringsmetodik för miljöpåverkan*, D57:2001, Licentiate thesis, Gothenburg: Chalmers University of Technology (in Swedish).

Matsson, L.O. (1999) *LCC-jämförelser CAV, VAV och ERIC*, Solna: Theorells Installationskonsult AB (in Swedish).

Milton, D.K., Glencross, P.M. & Walters, M.D. (2000) Risk of sick leave associated with outdoor air supply rate, humidification and occupant complaints, *Indoor Air*, **10**, pp. 212–21.

Sodec, F. (1999) Economic viability of cooling ceiling systems, *Energy and Buildings*, **30**, pp. 195–201.

Sterner, E. (2000) Life cycle costing and its use in the Swedish building sector, *Building Research & Information*, **28**, pp. 387–93.

Stetiu, C. (1999) Energy and peak power savings potential of radiant cooling systems in US commercial buildings, *Energy and Buildings*, **30**, pp. 127–38.

Sveriges verkstadsindustrier (1996) *ENEU 94*, Stockholm: Sveriges verkstadsindustrier (in Swedish).

Vik, T.A. (2001) *Natural versus mechanical ventilation – A comparison of life-cycle cost aspects of different ventilation systems*, Unpublished research project plan, Trondheim: Norwegian University of Science and Technology: http://www.ntnu.no/arkitekt/bt/ansatte/vikdrplan.html, [1 Sept. 2001].

Wargocki, P., Wyon, D.P., Sundell, J., Clausen, G. & Fanger, O. (2000) The effects of outdoor air supply rate in an office on perceived air quality, sick building syndrome (SBS) symptoms and productivity, *Indoor Air*, **10**, pp. 222–36.

Weschler, C.J. & Shields, H.C. (2000) The influence of ventilation on reactions among indoor pollutants: Modelling and experimental observations, *Indoor Air*, **10**, pp. 92–100.

Yang, H., Burnett, J., Lau, K. & Lu, L. (2001) Comparing central and split air-conditioning systems, *ASHRAE Journal*, **43**, *5*, pp. 36–38.

Chapter 6
Performance Indicators as a Tool for Decisions in the Construction Process

Veronica Yverås

Introduction

Many of the first decisions made during design are of great importance, as they will influence the final product in terms of, for example, durability and energy efficiency. Unfortunately, these issues are not always taken into consideration for a variety of reasons, which leaves the customer with a product that fails to satisfy. A major problem is that performance requirements are rarely stated, making it extremely difficult to achieve good decisions during the early stages of design.

A performance indicator tool would force the designer to investigate and specify the requirements. It would also provide guidance to the designer during design in order to predict performance. Such indicators would provide the designers with a first rough estimate and help to present, and make the client aware of, the consequences of a decision.

The tool proposed by the research reported in this chapter aims to raise the status of building physics, at present a discipline with an understated role in the construction process. By giving higher visibility to building physics, it is believed that the risk of performance failures due to simple, yet all-too common, mistakes during design will be avoided.

State-of-the-art review

Decision support during design

Normally, when a decision has to be made, relevant information is collected and reviewed in order to reach an outcome according to the criteria set. In the early stages of design, it involves selecting system solutions that will lay the foundation for the resulting building performance. For various reasons, there are difficulties in acquiring, studying, applying and distributing information.

The concept of performance, which is key to the design phase, was created at the beginning of the 1960s. It was heralded as a new and fresh concept that was seen to be the means for realising the goal of industrialising the construction sector. By converting prescriptive building regulations to those that were performance-based, it was hoped to stimulate and improve the ability to innovate. Difficulties in applying the concept were soon discovered; for example, implementation was slowed down as the older prescriptive codes were understood and therefore easier to use (Knocke 1970). Little or no regard was taken of the need for greater knowledge than had been the custom in earlier times. Designers had to be able to assess new technical solutions without tools or supplementary education and training. Over the intervening years, there has been some progress, but problems remain.

One of the main reasons for such slow progress is clients' inability to specify needs according to the performance concept. The client is often in a weak position relative to the designer, such that quality issues are not handled well. Unless the client demands better knowledge and communication skills, the sector will continue to mishandle these issues. Advances in ICT (information and communications technology), among other things, have spurred the development of visualisation techniques, that help to overcome many of the difficulties described above. By improving the quality of information during the design process, the client is better equipped to understand the different issues implicated in the project (Barrett & Stanley 1999).

Various authorities within Sweden have identified lack of knowledge during the construction process as one of the primary causes of sick buildings. Some initiatives have taken place. For example, the National Institute of Public Health has proposed a programme for improving the construction process. It offers guidance on how to engage any discipline that can contribute its knowledge to achieving a healthier indoor climate; for example, consultants with knowledge of allergies. The City of Stockholm has the intention of introducing requirements that will prevent damage due to moisture. Such action has been proposed before, but it was avoided as there was a strong belief that it would increase building costs. A corollary to this is that the Swedish government has recently established *The Quality Council of the Building Industry,* a body that brings together the different actors within the construction process with the aim of preventing failures.

Knowledge of building physics is a core competence of the Swedish National Testing and Research Institute (SP) in its quality labelling of healthy indoor environments. Adoption of their method requires that an expert is involved throughout the construction process, by contributing theoretical and practical knowledge (Carlsson 1999). The expert acts as a bridge between the theory of building physics and practice. All building projects successfully using this method receive the *P-label*, a quality sign. The method implies that designers on their own are not equipped to handle these issues and that there is a need to improve the conditions for making better decisions.

The use of the internet has also penetrated the field of building physics. CONNET and I-SEEC are examples of internet sites to support construction sector networking, making it possible for the digital dissemination of technical and product oriented information (Amor *et al.* 2001). An example Swedish internet site is that of Building Market (Byggtorget), that provides access to a wealth of knowledge on building related matters. That said, it is not clear if these sites have increased the search for available knowledge with the aim of making better decisions. Reliance upon familiar ways of working reduces the need to seek for information – a task that many actors in the construction process are unwilling to undertake (Davidson 2001). In the end, acquired information must be actively compiled and interpreted before a decision can be made.

The Construction Primer (Burry *et al.* 2001) is an on-line tool with a ready-made database for decisions created from a case-based reasoning. This involves using solutions from earlier decisions involving a similar problem (Schmidt-Beltz 1996). As the performance approach is not central to this method, it can cause difficulty when the results to be documented have to be measured.

The Hong Kong Building Environmental Assessment Method (HK-BEAM) is an initiative to encourage the actors in the construction process to strive for sound environmental performance in buildings (Burnett *et al.* 1998). By relating quantified performance to factors concerning the indoor environment, the building attracts credits for those criteria that are met. The building is then graded as fair, good, very good or excellent.

There is also criticism regarding the fragmentation of the sector in terms of its impeding the satisfactory formulation of performance requirements during the design process. One suggested solution is to educate building engineers in matters pertaining to building construction and building technologies, where performance reasoning is a key element (Carmeliet *et al.* 2001). This suggestion adds further weight to the argument for more attention being paid to the design stage.

Tools related to physical aspects

A conclusion that is often made, and which is incorrect, is that the complexity of theory is low because the information density is low (Mahdavi 1999). The inherent and complex physical relationships remain regardless of whether one is at the beginning or the end of the design process. It is nonetheless possible to consider the design process as divided into two phases, with a first-principle

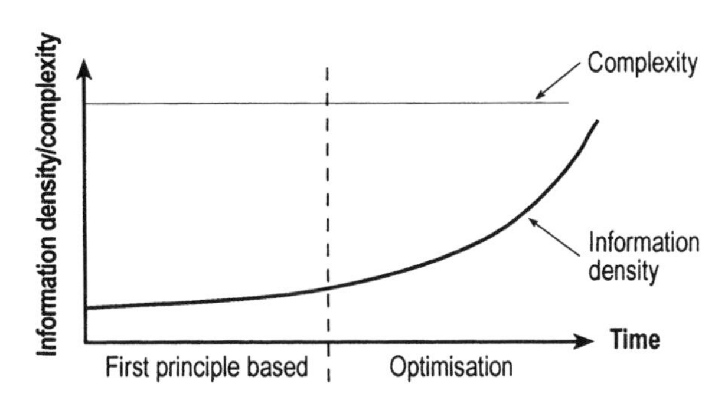

Fig. 6.1 Structure of the design phase.

based analysis quickly transforming into an optimisation process (Fig. 6.1). Decisions made in the first part of the design phase involve selecting a system by performance assessment. This analysis should be integrated in the project's programming phase as some decisions made there (or later) can lead to the use of solutions that are less satisfactory from a performance perspective. At the end of the design phase the optimisation of selected solutions takes place, reinforcing the reliability of the initial estimates.

Much research and development has been undertaken to produce simulation tools for use by designers during the optimisation phase. The extent of tools is so large that there is an organisation, International Building Performance Simulation Association (IBPSA) aimed at the development of simulation tools for application within the construction process. Despite the wide range of tools available, their quantity is not in line with their actual use. One cause is believed to be that some tools can assess only individual performance characteristics; for example, heat flows. Another cause is thought to be that they require expert knowledge when used (Hendriks & Hens 2000).

A decision support system is being developed to present the results of research into lay terminology. It identifies the relationships between factors that have direct and indirect consequences for the healthiness of the built environment and indoor air quality (IAQ) (Phipps *et al.* 2001). The aim of the system is to provide building designers, who are not IAQ experts, with a tool to make correct decisions. An inbuilt assumption is that IAQ problems occur when four major elements are present – source, pathway, driving force and occupants – and these form the core of the conceptual model on which the system is based. However, the system has been designed for New Zealand conditions only, bringing into question its wider application. Nonetheless, the system may have use in other regions, where conditions are similar.

Moisture measurement is a topic of particular concern and for which a method has been developed that can be handled by different users with varying levels of knowledge. The method assesses moisture protection, which has a strong influence on durability (Fig. 6.2). Expert knowledge is required for

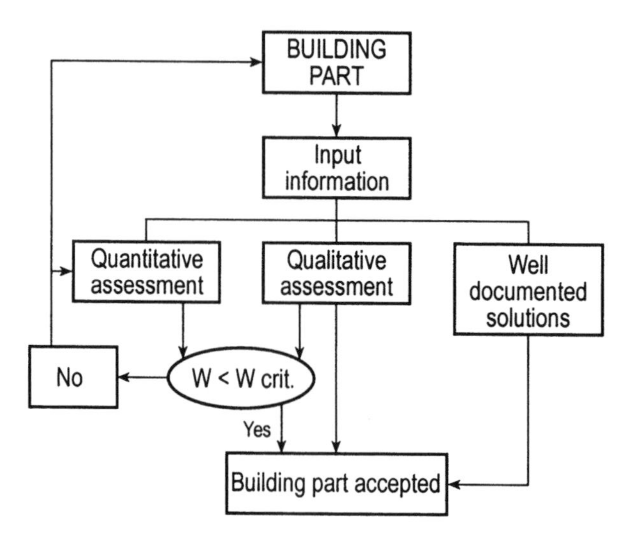

Fig. 6.2 Model for moisture measurement.

quantitative assessment, based on detailed calculations, in order to evaluate proposed technical (i.e. design) solutions. In this qualitative assessment, simple tools such as tables and diagrams are used to predict moisture conditions (Harderup 1999). However, an algorithm using performance indicators offers a simpler procedure for performance assessment. For those who are aiming at using a reliable solution that is well documented and for well-known conditions, case-based reasoning will suffice. A weakness of this approach is, however, that it can only be used when most decisions have been taken, since the starting point is the drawings. Changing decisions at this stage involves additional expense.

Buildings today have become so complex that it is no longer acceptable to assess the performance of individual aspects apart from each other, as they are interdependent to varying extents. Air tightness is of importance for energy efficiency, but demands appropriate ventilation in order to achieve good indoor quality. Hence tools have been developed with a holistic approach in mind and where the end-users are an important ingredient. *Integral building envelope performance assessment* (Hendrik & Hens 2000) adopts a holistic approach, in contrast to most other tools that consider the performance measure of energy efficiency. The *multiple criteria decision aid procedure* (Azar & Hauglustaine 2001) adopts a multi-actor and a multi-criteria approach. A *shared conceptual model* for integration in the building envelope design process that improves the transfer of data throughout design has also been proposed (Bédard *et al.* 1999).

The above tools aim at facilitating communication between actors. It is important to stress that the communication tools described assume that the actors within the design process have the necessary knowledge and experience. Lack of experience is a weakness that has to be dealt with consciously.

In order to assess an aspect of performance it is important to understand the interaction between different performance characteristics and their underlying theories. But even if knowledge is present, it is difficult to apply in practice in a systematic way without supporting tools. Aygun *et al.* (2001) present a parametric conceptual model for generating and evaluating various alternatives for a building. The model is divided into three subsystems: constraints (conditions of the environment), performance requirements and an element model. By displaying the subsystems in tables, the interaction of requirements and components (element) and the adjacencies of components are determined and presented. Performance requirements can therefore be expressed by the function of relevant instances of the three subsystems.

Performance indicators as tools

Indicators are often used within several areas. It is a powerful tool that is able to produce a summarised and easy-to-grasp picture of an often-complex situation. A tool may be used as a decision support system in different circumstances and also as an evaluation method. In an analogous case, the economic conditions in a country can be presented by, for example, gross domestic product (GDP), inflation and level of unemployment. As such indicators are internationally well established, they can be used to compare countries. Industries also use indicators as a tool for measuring, for example, productivity and environmental performance.

An area closer to home is that of investigative methods relating to sick buildings. The starting point has, in many cases, been technical measurements that have involved not-so-well-considered and often-expensive methods. Prevailing technical measuring methods have not always clarified the cause of the problem. In the regional hospital of Örebro in Sweden, an investigation strategy was developed that did not have technical measurement as its starting point (Andersson 2000). A questionnaire was used to map out experiences of the indoor climate. The results of the questionnaire were gathered and presented graphically in Fig. 6.3. By using this method, a first-principle-based analysis and guidance for further investigations have been achieved. In this example, the peaks in the diagram indicate a problem with the ventilation system. This is a resource and cost-efficient way of analysing possible causes of a problem by relying on environmental factors and symptoms only.

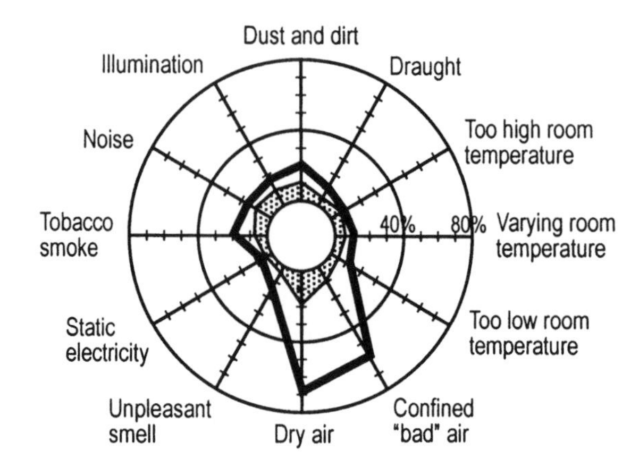

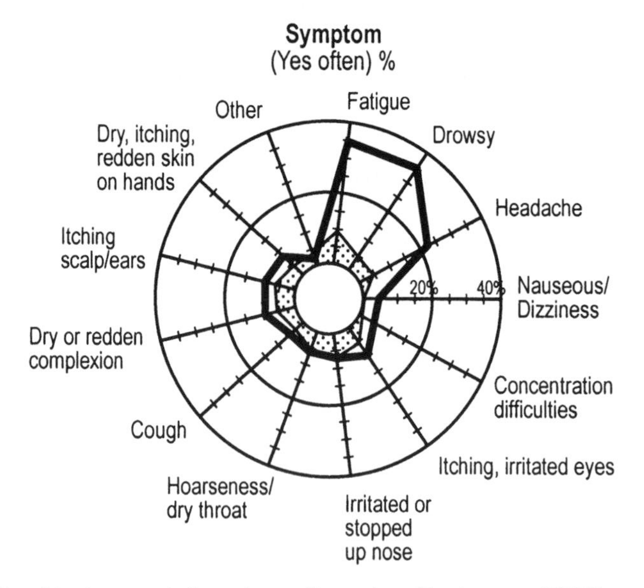

Fig. 6.3 Graphical presentation of questionnaires (Andersson 2000).

Research project

Project description and objectives

The objective of the proposed research is to create a decision support system that can be used in the early stages of design. Performance assessment is achieved by adopting a first-principle-based analysis to provide decision support and guidance for further design activity. By using performance indicators, information can be presented such that inexpert clients can understand the output of selected performance requirements and system solutions. The indoor environment, thermal comfort, energy efficiency and durability will be predicted by systematically assessing the performance of the building.

Research methodology

Indicators have been used successfully in several areas to describe and interpret rather complicated relations. They generally provide a simplified picture of reality, but are reliable enough for further decisions. Performance indicators are therefore applied, as the aim is to simplify decision-making. In order to assess the different performances of a building systematically, by using performance indicators, an algorithm of the design process has been adopted – see Fig. 6.4. By applying this algorithm, the designer and the client have to decide upon the performance requirements early in the process, unlike the model for moisture measurement shown in Fig. 6.2.

The identification and listing of performance indicators that are the main ingredients of the decision tool can be undertaken in different ways. One alternative is to analyse cases (good and bad), but it would not be a rational or a systematic method; moreover, it would require many cases. A plausible view is that the theory should be the starting point as it can provide the knowledge needed to design healthy buildings. The extent of available literature and the complex mechanisms of building physics make it essential to classify and link the knowledge appropriately according to a fixed model. Flowcharts showing prevailing stresses are a way of identifying relevant indicators. The stresses

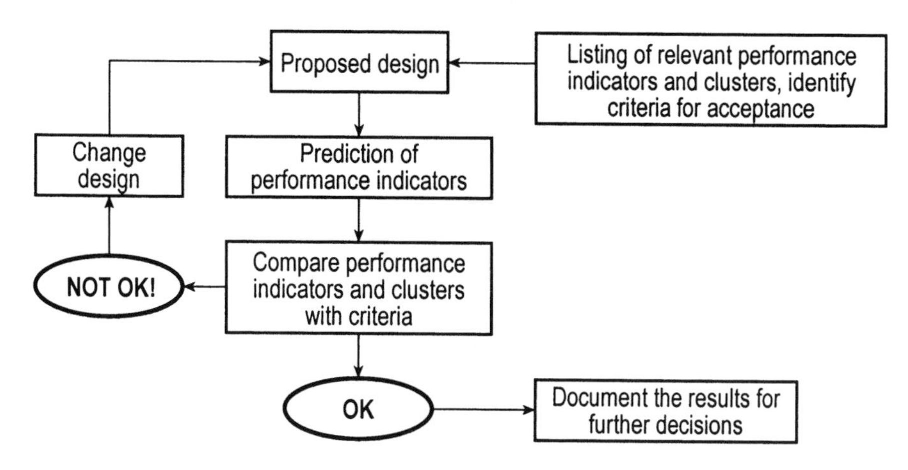

Fig. 6.4 The algorithm of the design process.

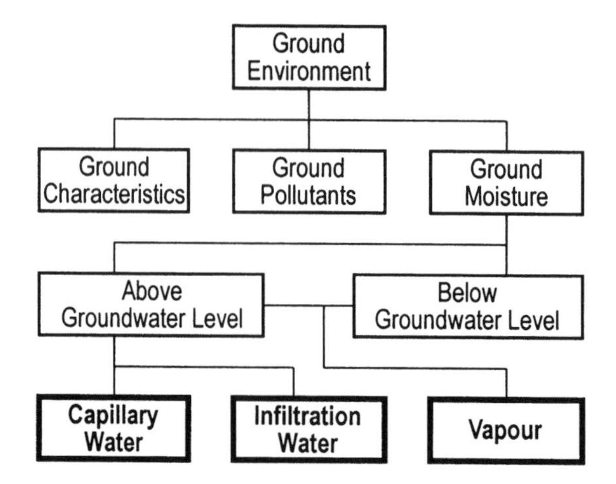

Fig. 6.5 Ground environmental stresses.

are organised under four main headings: *outdoor environment, indoor activities, system* of the building and *ground environment* (Fig. 6.5).

When a stress cannot be divided further, a performance indicator flowchart is used. It describes the theory in terms of stress-mechanism-output-performance requirement – performance indicators (Fig. 6.6). This way of describing a process has been performed in a similar way elsewhere, where the degradation process of a material has been described in terms of *agents-mechanism-effect* (Norén 2001).

Performance indicators need to be predicted in order to provide a value that is needed for comparison with the chosen criteria for acceptance. This can be done in several ways but the aim, as stated before, is to keep it as simple as possible. Prediction methods are, therefore, being identified through the

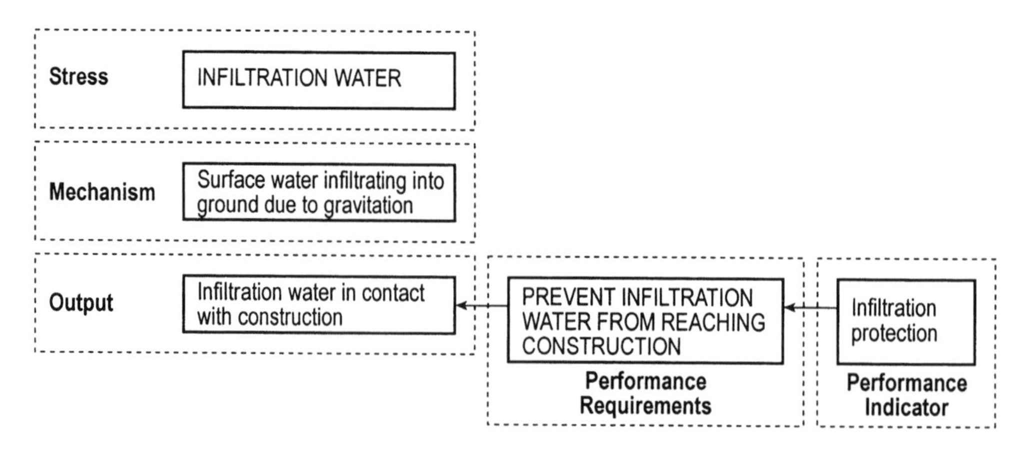

Fig. 6.6 Flowchart model of infiltration water stress.

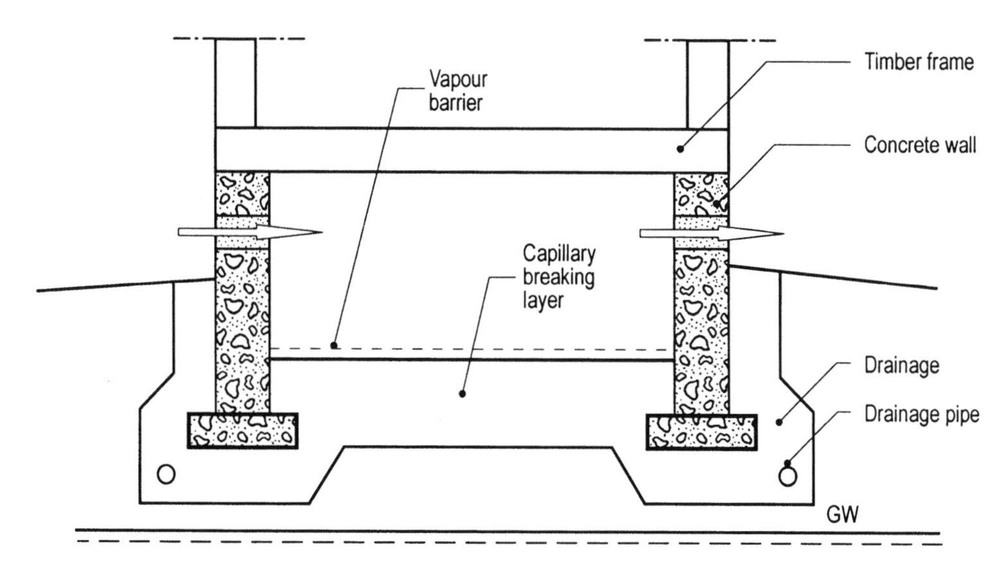

Fig. 6.7 Crawl space.

use of diagrams, checklists, tables, rules of thumbs and simplified calculation methods. A simplified case study has been carried out in order to identify an appropriate method of assessing a known technical or system solution. The case is an outdoor ventilated crawl space with wooden beams – see Fig. 6.7. The evaluation of this case study, in terms of reliability, was achieved by literature studies and computer simulations.

Research results and industrial impact

Quantification of results

In the case study, four relevant performance indicators were identified, as shown in Table 6.1. The vapour protection indicator is a result of two other performance indicators that consider the vapour barrier and the evacuation of vapour. These two indicators are weighted together into one – *the vapour protection*. Lastly, the durability of ground construction has to be assessed. In this case, indicators (I) and (III) in Table 6.1 are regarded as checklist items since they are both considered to achieve acceptance degree 1 and are therefore not a problem of durability. However, indicators (II) and (IV) are the two having the greatest influence on durability and are therefore regarded as the main indicators in this assessment (Fig. 6.8).

Durability is displayed as arrows in the diagram to indicate the stress load, which reduces the performance of the crawl space. In other words the shorter the arrows are, the better. If the diagram shows a long arrow, the crawl space

Table 6.1 Relevant indicators when assessing the durability of an outdoor ventilated crawl space with different levels of acceptance.

Stress	Performance indicator	Performance requirement	Acceptance (A-degree)
Ground Moisture (see Fig. 6.2)	*I. Capillary protection*	Prevent capillary water from reaching the construction	The construction will not get into contact with capillary water No protection
	II. Vapour protection	Prevent ground vapour from increasing the moisture content in construction	A-1= Vapour from the ground is excluded A-2= Vapour stress is partly reduced A-3= No vapour protection
	III. Infiltration water protection (Fig. 6.3)	Prevent capillary water from reaching the construction	Liquid water is excluded Liquid water expected
Air Moisture	*IV. Cooling reduction of air*	Reduce the cooling effect of air (outdoor ventilation) in construction	Reduced cooling effect No reduction is achieved

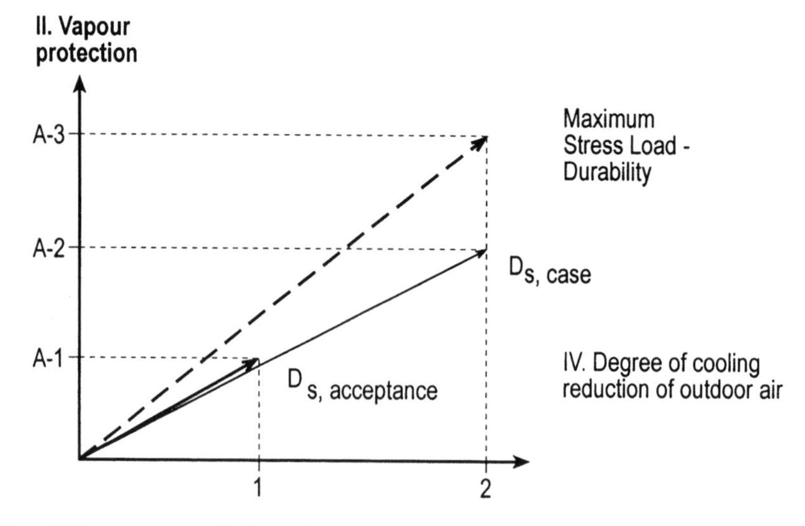

Fig. 6.8 Durability assessment of crawl space.

will suffer a high stress-load. In our case, the degree of acceptance is rather limited due to the selected material (wood) in the construction and therefore leads to the low tolerance of $D_{s,\ acceptance}$. As illustrated, $D_{s,acceptance} < D_{s,case}$ which is not acceptable (Fig. 6.4). This means that the construction needs to be redesigned or another technical solution must be considered.

Implementation and exploitation

The performance indicator tool, as outlined, effectively constitutes a decision support system that will facilitate making the right decisions from the start. There are many examples of how poor information can affect the final outcomes from the construction process. One of the more recent is the Museum of Modern Art in Stockholm, where the client had to move out not long after opening because of the incidence of sick building syndrome (Kyander 2001). By adopting performance indicators it will be possible to predict the likely performance of a building and thereby minimise its risk of failure. The tool has the potential to support the designer when considering relevant stresses and also in explaining the different mechanisms that control different building physics processes. In a longer-term context, it means safer design and a lower risk exposure for the companies concerned. Communication between designer and client will be facilitated, due to being able to illustrate the output of a decision. A classic example is when the client wants to adopt a shorter construction phase that does not always correspond to the demands of the materials selected; for instance, time for drying out.

Recently, Swedish environmental law has been amended. The burden of proof has been reversed so that owners of buildings have to provide evidence that the regulations have been followed if a tenant complains about the indoor environment. Another change involves businesses having a duty to acquire relevant knowledge – the role of the informed client. In practice, this means that the client is responsible for seeing that designers and contractors have the necessary knowledge to ensure a good indoor environment. In this context, performance indicators will be an aid to having all relevant facts to hand in order to make good decisions.

Conclusions

Simplicity has been central to the development of the tool, since designers today suffer under the pressure of too little time in which to make important design decisions. Gaps in basic knowledge are common and so it would be pointless developing and deploying complicated, specialist tools. It is also unreasonable to begin design with the use of specialist simulation tools, as the first step has to be to outline the design. By using performance indicators, the designer will be guided in a straightforward manner to identify and consider different stresses to which a building or parts thereof might be exposed. This would reduce the risk of mistakes occurring during the design phase.

The proposed algorithm forces the actors involved in the design to describe the performance requirements as the criteria for acceptance as early as possible. The minimal application of performance concepts during decision-making nowadays is, however, an issue that has to be faced (Becker 1999). There has to be a change concerning the attitude towards spending time assessing the performance of the building. Since prescriptive building codes were replaced, the design process has changed little in adjusting to the new concept. Another issue is the determination of performance requirements, since it is difficult to specify the needs of users. This is an area that has to be investigated further.

The case study shows how durability is affected by the design. The proposed tool provides the designer with guidance when outlining the design. This provides a first analysis of the performance and might suggest that a further deeper analysis is necessary. The tool does not produce an exact answer, but it gives an idea of how the building part or a part of it will perform.

References

Amor, R., Bloomfield, D., Groosman, M. & Rio, O. (2001) Information services to enable European construction enterprises an overview of the I-SEEC European Union project, *CIB World Building Congress, Wellington, New Zealand, 2–6 April 2001*, pp. 4/31–41.

Andersson, K. (2000) Kan man från mätresultaten uttala sig om hälsorisker? *Inomhusklimat, Örebro, Sweden, March 14–15 2000*, pp. 74–80 (in Swedish).

Aygun, M., Cetiner, I. & Gocer, C. (2001) Specification of performance requirements for building elements by conceptual models, *CIB World Building Congress, Wellington, New Zealand, 2–6 April 2001*, pp. 2/133–137.

Azar, S. & Hauglustaine, J-M. (2001) Interactive tool aiding to optimise the building envelope during sketch design, *Building simulation '01 conference*, Rio de Janeiro, Brazil, 13–15 August, IBPSA, pp. 387–94.

Barrett, P. & Stanley, C. (1999) *Better construction briefing*, Oxford: Blackwell Science.

Becker, R. (1999) Research and development needs for better implementation of the performance concept in building, *Automation in Construction*, **8**, 4, pp. 525–32.

Bédard, C., Fazio, P., Ha, P. & Rivard, H. (1999) Shared conceptual model for the building envelope design process, *Building and Environment*, **34**, 2, pp. 175–87.

Burnett, J., Jones, P., Lee, W.L. & Yik, F.W.H. (1998) Energy performance criteria in the Hong Kong Building Environmental Assessment Method, *Energy and Buildings*, **27**, 2, pp. 207–19.

Burry, M., Coulson, J., Preston, J. & Rutherford, E. (2001) Computer-aided design decision support: interfacing knowledge and information, *Automation in Construction*, **10**, 2, pp. 203–15.

Carlsson, M. (1999) Kvalitetssäkring av god innemiljö, *Provning & Forskning*, 1, p.12 (in Swedish).

Carmeliet, J., Hens, H., Janssens, A. & Vermeir, G. (2001) Building engineering – a needed discipline, *Thermal Envelope and Building Science*, **24**, *4*, pp. 266–74.

Davidson, C. (2001) Performance based building – from theory to practice through information and knowledge management, *CIB World Building Congress, Wellington, New Zealand, 2–6 April 2001*, pp. 1-53.

Harderup, E. (1999) *Fuktdimensionering med generell checklista,* Report TVBH-3031, Lund: Lund Institute of Technology (in Swedish).

Hendriks, L. & Hens, H. (2000) *Building envelopes in a holistic perspective,* IEA Annex 32, Task A, Leuven: Laboratorium Bouwfysica.

Knocke, J. (1970) *En funktionsanalytisk byggnorm,* R21:1970, Stockholm: Statens institut för byggnadsforskning (in Swedish).

Kyander, K. (2001) Husvillt museum, *Arkitekten*, **10**, pp. 18–20 (in Swedish).

Mahdavi, A. (1999) A comprehensive computational environment for performance based reasoning in building design and evaluation, *Automation in Construction*, **8**, *4*, pp. 427–35.

Norén, J. (2001) *Assessment and Mapping of Environmental Degradation Factors in Outdoor Applications*, Stockholm: Royal Institute of Technology.

Phipps, R., Laird, I. & Wall, G. (2001) Decision support for healthy office building design: a conceptual framework, *CIB World Building Congress, Wellington, New Zealand, 2–6 April 2001*, pp. 1/495–503.

Schmidt-Beltz, B. (1996) Scenario of an integrated design support for architects, *Design Studies*, **17**, *4*, pp. 489–509.

Chapter 7

Reducing the Risk of Failure in Performance within Buildings

Stephen Burke

Introduction

Building failures occur daily because of misunderstandings over the importance of building physics, especially among the different actors in the construction process, each of whom may have a different appreciation of causes and effects. Legislation, which has been passed to ensure the health and safety of the occupants, has addressed some of the larger performance issues such as thermal comfort, energy usage and, to some extent, indoor air quality. However, other types of building failure continue to be ignored. These are the longer-term soft issues such as high moisture content in a building, high-energy costs and the overall sustainability of the building.

These types of building failures are believed to be linked to health problems and are largely preventable with today's knowledge of building physics. This chapter looks at current building failures directly attributable to the neglect of building physics principles, and why it is important to include these factors actively in design decision-making.

State-of-the-art review

Building physics is comprised of many various components. This section will look at these components, the role that economics plays and the effects that building failures have on people's health.

Building physics

Building physics is the science of how matter and energy interact within a building system. More specifically, this field encompasses the areas of heat, air (ventilation), moisture flows and the energy interactions between all of them.

It is important to note that the Swedish context of building physics does not include acoustics and fire protection as in other countries (Sandin 1990). This area of science exists to ensure that people have an area to live in that provides thermal comfort and does not cause health problems. Many health issues arise when fundamental physical principles are overlooked or ignored and this can translate into higher costs for society as a whole.

The literature related to economic aspects of building physics is negligible. Jóhannesson & Levin (1998) attempted to examine these two areas concurrently in their paper by looking at the relationship between design that neglects common theories of building physics and the consequent environmental and economic cost. In their paper, a typical Swedish single-family dwelling was examined under two scenarios: one without any special environmentally friendly materials or features, and the other incorporating materials that are considered to be environmentally friendly by current standards. The authors focused on the performance of available environmentally superior materials when considered in a life cycle cost (LCC) context and found many to be inferior. Using more environmentally friendly materials did yield lower energy use and costs for construction. Over the operational life, the difference in performance between the two buildings was great enough for the environmentally friendlier building to be more energy-intensive over its life than the standard building. The weakness in this paper, as declared by the authors, is that it does not look at the economic repercussions of either including or excluding building physics issues in themselves; but it is able to give an idea of the result of ignoring building physics principles.

It appears that most of the literature that is related to building physics looks at building physics in its various components and does not bring economic factors into view. The literature generally considers potential problems that a building can have from a health perspective and tries to attribute the problem to one or two faulty components in the building and their possible remedy.

Heat flows

Heat flow principles are the backbone of modern, energy-efficient buildings. In order to make a building use less energy for heating and cooling, walls are insulated with materials that minimise the heat flow between the inside and outside of a building. Another benefit of an energy-efficient building is the level of thermal comfort that people feel in the building. People are very sensitive to temperature changes and even a small heat loss is detectable. Thermal comfort is not, however, determined by temperature difference alone; moisture levels and air pressure can also have an effect on people's thermal comfort.

With energy-efficient buildings come new problems in the areas of moisture and ventilation. While heat flows are in themselves largely understood, the repercussions of energy-efficient buildings on moisture levels and health are not. As a building's energy efficiency increases, so does its potential for moisture and ventilation problems if it is not designed properly (Thörn 1999). The design, construction, materials and workmanship of the smaller components of a building become more important and can make a significant difference in building performance. For example, reducing the thermal bridging in a typical Canadian timber-framed wall by applying an exterior insulated sheathing yields a 12% gain in efficiency (Ministry of Housing 1990).

Europe is currently trying to find a way of encouraging people to build more energy-efficient homes. One means of accomplishing this is by creating an

> 'energy certification … [offering] advice for new and existing buildings and a public display of certificates in certain cases' (ECCP 2001).

In addition to this, new and newly renovated buildings will have to meet a minimum standard of energy performance, and this will be measured by a standardised measuring system. This should force designers and construction companies to increase their level of performance, resulting in higher quality buildings than we have today.

The United States has also initiated a similar programme called High-Performance Commercial Buildings: A Technology Roadmap.

> 'The fundamental goal is to optimise the building's performance in terms of comfort, functionality, energy efficiency, resource efficiency, economic return, and life-cycle value' (Swartz 2001).

This plan will be executed over the next 20 years and will involve the US Department of Energy (DOE) and its partners in both the public and private sectors.

It is becoming more common today for researchers to combine heat and moisture flows in their research areas and attempt to answer the questions of what effects heat flows in buildings have on moisture flows and ventilation requirements (Samuelson 1998). The theory behind moisture and heat flows are very similar, but there is one important factor that makes moisture calculations much more complex than heat flows. The difference is that when heat flows are calculated, the materials are assumed to be dry and the effects of thermal conductivity are negligible. Calculations are therefore based on a constant temperature and moisture state. Yet, moisture properties are very sensitive to changes in both temperature (the vapour permeability is temperature dependent) and moisture state. One cannot assume a constant temperature or moisture state when calculating moisture flows. In addition, other factors such as the material's properties and air velocity can have effects on the flows.

A careful balancing act is required to obtain buildings that are both energy-efficient and healthy for its occupants (Sandin 1990).

Computer software is available for calculating heat flows in buildings. Most of these are specific programs designed to calculate the heat flows of various components like attics, crawl spaces and walls. Some of the software such as *MOIST, HEAT2* and *HEAT3* look at the one, two and three-dimensional steady state of a design respectively. The user is able to improve the heat resistance of the design by changing both the design and materials used in the simulation in order to decrease the amount of heat energy lost from the building. The main concern with current software that is available today is that it is difficult to use (Blomberg 2000; Burch & Chi 1997).

There are also many computer programs available for calculating the energy usage of a building. These programs usually take into account the weather, type of windows, type of walls, and other specific details of a building. Each program usually has a feature that sets it apart from others. For example, *ENERGY-10* can include passive solar heating, glazing and thermal mass in the design phase (EREC 2000). *NHER Evaluator* also calculates energy usage; however, it has the option of calculating surface condensation and the effects of cold snaps (NES 2000). Using software of this sort enables users to test alternative materials and designs in order to optimise their building's energy usage.

Air flows

Ventilation is the link between the indoor air and outdoor air of a building. With proper ventilation, a building has a readily available supply of fresh air that keeps the interior thermal environment comfortable and moisture levels under control. As buildings become more energy-efficient, they are required to be more airtight. This places more importance on a properly designed and balanced ventilation system. Without properly designed ventilation systems, buildings can rapidly become odorous and unhealthy to the occupants, because of a build-up of chemicals, moisture and organic compounds. In recent years there have been many studies looking at the relationship between indoor air quality and ventilation rates. The consensus is that, up to a certain point, the lower the ventilation rate, the worse the indoor air quality (Fisk 2000; Sundell 2000; Wargocki *et al.* 2000; Apte *et al.* 2000; Milton *et al.* 2000). For a more detailed examination of the importance of ventilation systems, see Chapter 5 (*A life cycle cost approach to optimising indoor climate systems*).

Moisture flows

Moisture and its effects on a building is a common topic of discussion within the building physics field today (Luthander 2001; Samuelsson 2001). The topic is, however, very complex and covers many areas of science.

If we want to understand fully the problems caused by moisture we must look to various disciplines in the scientific community. They include medical doctors and researchers, microbiologists, biologists, physicists, chemists and engineers. By taking a multidisciplinary approach, we can begin to understand the nature of the problems associated with moisture in buildings and how to prevent them (Wolkoff *et al.* 1997; Sundell 2000).

Three methods of transportation that enable moisture to come into contact with materials are convection, diffusion and capillary action. Convection processes involve moving air that picks up and deposits moisture on the surface of materials. Diffusion of moisture through the air contributes less moisture to a material than convection, due to the volume of moist air that is exposed to the surface of the materials. Capillary action mostly takes place underground, when groundwater is drawn into the materials (Sundell 2000; Nevander & Elmarsson 1994).

There are many different paths for moisture to enter a building and these can appear during any of the different stages in the construction process. Before the construction phase, some of the materials can be shipped wet to the job site. For whatever reason, these materials are not allowed to dry properly or they are exposed to water in storage or during transportation. Even if materials are shipped dry, they sometimes become wet at the job site due to improper storage. If materials are stored properly, i.e. stored indoors or covered up, the risk of a building becoming damaged due to moisture can be significantly reduced (Sundell 2000; Nevander & Elmarsson 1994).

After the building is complete, there is still a risk of damage from both the indoor and outdoor environments. Figure 7.1 shows some of the possible damages that can occur when different materials are exposed to various levels of moisture. The majority of moisture damage begins once the relative humidity has reached a level of around 75%. This shows the importance of proper ventilation and indoor climate control. Indoors, people contribute to the moisture level by physically sweating, cooking food, taking showers etc. In other buildings such as paper mills, swimming pools, and other facilities that use a lot of water, there is a very high risk of moisture damage due to condensation (Sundell 2000; Nevander & Elmarsson 1994).

The greatest outdoor risk to a completed building is the weather. Rain, snow and humid air can result in exposed materials becoming very moist. In addition, leaks in the vapour barrier and in the roof can allow moisture to come into contact and contaminate various materials.

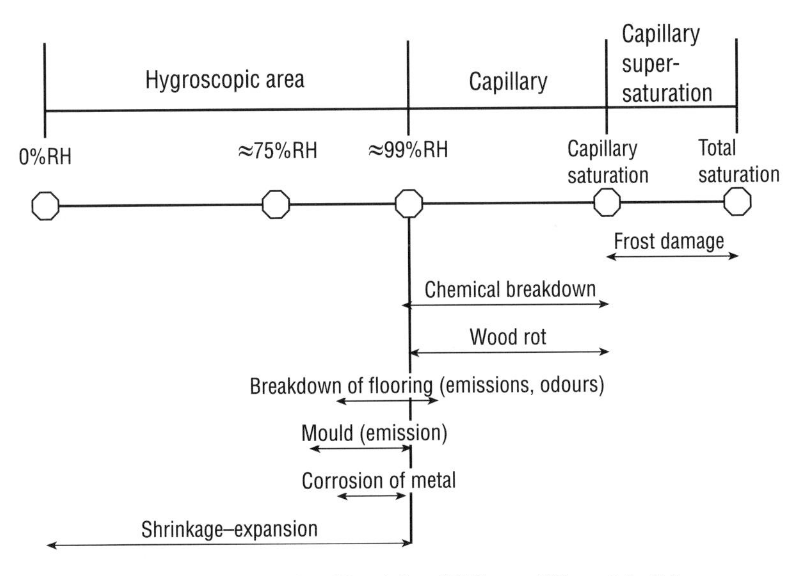

Fig. 7.1 The effects of Relative Humidity (RH) on different building components (based on ideas, contributed during a discussion, by Professor Göran Fagerlund, Lund Institute of Technology, Lund, Sweden).

Despite the availability of this knowledge, the same problems continue to occur. A recent, large-scale housing project (Hammarby Sjöstad in Stockholm) that was designed according to environmentally friendly principles became a focus for the mass media when moisture problems developed in some near-finished apartments. The problem was attributed – though not proven – to an unusually moist summer during the on-site production phase. The materials were not protected adequately from the moisture and in turn became wet. Not long after the heating system was activated, mould began to grow on the materials, contaminating a number of the apartments (Luthander 2001). Problems of this nature are not limited to Sweden; Denmark, the UK and many other countries are seeing an increase in moisture-related problems. These projects illustrate that even well designed, environmentally friendly buildings can have moisture problems due to a combination of weather, materials and, increasingly, the on-site production process.

Computer programs that attempt to predict the likelihood of moisture penetration are increasing in number. However, they are few in comparison to heat modelling programs due to the complexity of moisture flow. In order to develop a reliable moisture model, an accurate temperature model must be used. Some software exists today: *WUFI* from Germany is able to calculate both heat and moisture flows in a one- or two-dimensional scale using different materials (Gertis 2000). *RISK1* from Sweden is a one-dimensional program that calculates the risk of moisture damage to a building depending on its geographic location (Harderup 1999); and *MOIST* from the US is a one-dimensional heat and moisture calculation program (Burch & Chi 1997).

Economic aspects

Failure in the performance of buildings results directly in financial loss for owners, occupants and other stakeholders each year. According to Fisk (2000), in the US, the estimated potential annual savings and productivity gains are US$ 6–14 billion from reduced respiratory disease, US$ 2–4 billion from reduced allergies and asthma, US$ 10–30 billion from reduced sick building syndrome symptoms, and US$ 20–160 billion from direct improvements in worker performance that are unrelated to health. In Sweden, about SEK 6 billion (€665 million) per year is spent on repairs and maintenance. Of that, roughly half goes to damages attributed to moisture damage (Tolstoy 1994).

Today, the average Swedish household can expect to pay SEK 200 000–300 000 (€22 000–33 000) to repair its moisture-damaged home. These costs do not include health care costs associated with asthma, multiple chemical sensitivity, sick building syndrome and reduced productivity, all of which can be caused by a building with poor performance. While it is debatable whether some of these illnesses can be attributed to physical or psychological causes, it remains the case that people suffer from symptoms caused by inadequate buildings (Willers *et al.* 1996; Wolkoff *et al.* 1997; Arnetz 1999; Terr 2000; Apte *et al.* 2000; Milton *et al.* 2000; Wargocki *et al.* 2000).

It is rare for construction companies – house builders in particular – to calculate and show their customers the long-term operating costs and likely problems associated with a specific design. Barrett & Stanley (1999) touch on the issue of customer empowerment, arguing that the construction sector should empower its customers and should enable them to become more acutely aware of the details of their project (or home). This could be interpreted as arguing the case for greater awareness on the part of customers. If customers are insufficiently experienced to address certain issues of the construction process, the sector should provide information to enable them to fulfil their role as informed (or intelligent) clients. This information could be in the form of possible and known design issues, estimated operating costs such as energy usage over the lifetime of the building, maintenance costs or the overall sustainability of the building (Barrett & Stanley 1999).

When a building is found to malfunction, the materials of its composition are among the first items to be investigated. The decision to use one material over another is usually more of an economic issue than one of performance. An exception to this is if a more expensive material is chosen for its environmental properties. However, this does not mean that the material is superior physically. Usually, greater emphasis is placed upon materials that are cheap and that will perform to the minimum performance required by legislation and building codes. This helps to keep the cost of the building down, which is an advantage to both the construction sector as well as the economy.

Recognising that the lowest bid does not always represent the most economically advantageous solution is a widely held view. Some countries have adopted other methods such as eliminating the highest and lowest bids and accepting the bid closest to the average (Hatush & Skitmore 1998). While this method can help to reduce short-term building failures by allowing companies with better quality control over workmanship to be awarded contracts, this does not guarantee that long-term building failures will be avoided. The operating costs of a high-quality building can still be high due to inadequate measures regarding energy use. However, companies can reduce the risk of both short and long-term problems and energy costs by addressing aspects of building physics early in the construction process.

Research project

Project description and objectives

The aim of this project is to determine and develop tools that could be used to include aspects of building physics in the design stage of buildings, by highlighting some of the economic benefits. This project will entail the identification of building physics areas having the largest potential impact on the decision-making process, the design of a tools package for use in the sector, the testing of this package by companies and the analysis of results. The main objective of this project is complementary to other projects discussed in this book. The concept behind this particular project is that by improving the quality of new buildings, the construction sector will gain techniques and knowledge that will make it more competitive on both the national and international level.

Research methodology

This project makes use of qualitative and quantitative methods. In the early days of this project, a literature study was conducted in order to determine the areas where the building physics tools package would have the most beneficial impact on the construction sector. This will be continued during the entire project. A problem with literature surveys is that they do not indicate the kinds of tools the sector might be interested in deploying.

Interviews and surveys are methods of enquiry that could be used to help in defining such tools. These methods would be more reliable than a literature study; however, such methods can be problematic, not least because of the difficulty of establishing a representative sample. That said, it is recognised that commercial exploitation of the results of the research would most probably require sampling of the population.

Participant observation is an alternative method that could be used to record a construction company's activities and enable a tools package to be designed to fit those activities. However, this was not considered to be an appropriate method, because of the amount of time required to conduct observations.

Once a tool is designed and developed, a few methods are available for analysis and verification. The primary method under consideration is a detailed case study of one company to determine what effects the tool has had on the company's decision-making process when planning a building. An alternative is to use the tool with a number of different companies and create examples that are less detailed, but perhaps more representative of the construction sector than a single case study. A later phase in the research may, therefore, include a set of case studies in which specific experiences of using the tools package might be investigated.

Research results and industrial impact

Quantification of results

In this project, we expect to see that the cost of actively incorporating building physics principles into the design phase of a building is demonstrably less than the potential consequences of not including them. Up to now, the literature suggests that there are large potential cost savings from integrating building physics principles into design decision-making. The literature also suggests that these savings should come primarily from reduced renovation costs, increased energy efficiency of the building during occupation, increased performance from workers and decreased sick leave/health care costs.

Implementation and exploitation

Discussions will take place with related projects under *Competitive Building*, with a view to having the resultant tool utilised in practice – see also Chapter 6. The aim will be to help designers achieve greater energy efficiency and avoid, or at least reduce substantially, moisture problems. A tools package of this nature could also increase the competitiveness level of Swedish construction companies on both the national and international level.

The incentive for utilising the tools package in routine design is to cut ownership costs for the customer (and taxpayers) by reducing the amount of repairs arising from moisture problems and health problems associated with the occupation of buildings. If customers were made aware of the tools available to designers and construction companies, they might force their use through demands for better information. There is also the health dimension. Operatives who feel healthier are more productive; even a small increase in productivity can translate into worthwhile profits for a company. In addition, by ensuring a high level of healthiness, the amount of health care resources needed may be reduced.

Conclusions

Today, a considerable amount of money is being spent on repairing buildings damaged by moisture, treating people who have become ill in their home or work environment and paying unnecessarily high energy bills. For the most part, technology exists to reduce these costs dramatically; however, the construction sector is generally not motivated to utilise the available information.

Building physics has the potential for builders to generate and for customers to save large sums of money each year. Easy-to-use tools for the sector are needed to highlight the economic benefit of key decision-making during design. This would help to create a smoother transition to designing better buildings than the present state of affairs.

References

Apte, M.G., Fisk, W.J. & Daisey, J.M. (2000) Associations between indoor CO_2 concentrations and sick building syndrome symptoms in U.S. office buildings: an analysis of the 1994–1996 BASE study data, *Indoor Air*, **10**, pp. 246–57.

Arnetz, B.B. (1999) Model development and research vision for the future of multiple chemical sensitivity, *Scandinavian Journal of Work and Environmental Health*, **25**, 6, pp. 569–73.

Barrett, P. & Stanley, C. (1999) *Better Construction Briefing*, Oxford: Blackwell Science.

Blomberg, T. (2000) *HEAT2 A PC-Program for heat transfer in two dimensions. Manual with brief theory and examples*, TVBH-7215, Lund: Lund Institute of Technology; Cambridge, MA: MIT Building Technology Group.

Burch, D.M. & Chi, J. (1997) *MOIST*, Washington: Government Printing Office.

ECCP (2001), European Climate Change Programme [Online], European Commission, Available from: http://europa.eu.int/comm/environment/climat/eccp.htm [27 June, 2001].

EREC (2000) Computer software programs for commercial/large building energy analysis, Energy Efficiency and Renewable Energy Network. Available from: http://www.eren.doe.gov/consumerinfo/v102.html [27 June, 2001].

Fisk, W.J. (2000) Review of health and productivity gains from better IEQ (Indoor Environmental Quality), *Proceedings of Healthy Buildings*, Helsinki, **4**, pp. 23–34.
Gertis, K. (2000) *WUFI*. Stuttgart, Germany: Fraunhofer Institut Bauphysik.
Harderup, L.-E. (1999), *RISK 1: Manual och teori (Manual and Theory)*, TVBH-3036, Lund: Lund Institute of Technology (in Swedish).
Hatush, Z. & Skitmore, M. (1998) Contractor selection using multicriteria utility theory: An additive model, *Building and Environment*, **33**, pp. 105-15.
Jóhannesson, G. & Levin, P. (1998) Building physics – no way around it, in *CIB World Building Congress Proceedings CD-ROM*, Gävle, 7–12 June, pp. 2–7.
Luthander, P. (2001) Skanskas byggslarv orsakade mögel (Skanska's construction carelessness caused mould), *Dagens Nyheter*, 13 February, A5 (in Swedish).
Milton, D.K., Glencross, P.M. & Walters, M.D. (2000) Risk of sick leave associated with outdoor air supply rate, humidification, and occupant complaints. *Indoor Air*, **10**, pp. 212–21.
Ministry of Housing (1990) *Code and construction guide for housing*, North York: Ontario New Home Warranty Program, Technical Research and Training Department.
NES (2000) The NES software catalogue, National Energy Services Ltd. Available from: http://www.nesltd.co.uk/software.htm [27 June, 2001].
Nevander, L.E. & Elmarsson, B. (1994) *Fukt handbok (Moisture handbook)*, 2nd edition, Stockholm: AB Svensk Byggtjänst (in Swedish).
Samuelson, I. (1998) CIB W40 Heat and moisture transfer in buildings – trends and future perspectives, *CIB W40 Heat and moisture in buildings*, Kyoto, Japan, 7–10 October, pp. 5–9.
Samuelsson, M. (2001) Moderna museet är mögelskadat (Modern Museum Damaged by Mould), *Svenska Dagbladet*, 14 June, p.2 (in Swedish).
Sandin, K. (1990) *Värme; Luftströmning; Fukt (Heat; Airflow; Moisture)*, 2nd edition, Lund: Lund Institute of Technology (in Swedish).
Sundell, J. (2000) Building related factors and health, *Healthy Buildings*, Helsinki, **1**, pp. 23–33.
Swartz, J. (ed.) (2001) *U.S. 20-year plan will lead to healthier, more energy-efficient commercial buildings*, Ottawa: Institute for Research in Construction.
Terr, A.I. (2000) Multiple chemical sensitivities, in Bardana E.J.J. & Montanaro, A. (ed.), *Indoor Air Pollution and Health*, New York: Marcel Dekker, pp. 269–278.
Thörn, Å. (1999) *The emergence and preservation of sick building syndrome – research challenges of a modern age disease*, Doctorial Thesis, Stockholm: Karolinska Institute.
Tolstoy, N. (1994) *The condition of buildings. Investigation methodology and applications*, TRITA-BYMA, Stockholm: Royal Institute of Technology.
Wargocki, P., Wyon, D.P., Sundell, J., Clausen, G. & Fanger, P.O. (2000) The effects of outdoor air supply rate in an office on perceived air quality, sick building syndrome (SBS) symptoms and productivity, *Indoor Air*, **10**, pp. 222–36.
Willers, S., Andersson, S., Andersson, R., Grantén, J., Sverdrup, C. & Rosell, L. (1996) Sick building syndrome symptoms among the staff in schools and kindergartens: are the levels of volatile organic compounds and carbon dioxide responsible, *Indoor Built Environment*, **5**, pp. 232–35.
Wolkoff, P., Clausen, P.A., Jensen, B., Nielsen, G.D. & Wilkins, C.K. (1997) Are we measuring the relevant indoor air pollutants? *Indoor Air*, **7**, pp. 92–106.

Chapter 8

Physical Status of Existing Buildings and their Components with the Emphasis on Future Emissions

Torbjörn Hall

Introduction

An important goal for the construction sector is to have the knowledge and tools that will enable it to satisfy, in an economically optimal way, the expected aesthetic, technical and environmental requirements of a building project. This must also include a healthy indoor climate for everyone. However, it is not always possible to satisfy these requirements.

In addition, in the case of refurbishment and rebuilding projects, there are often a greater number of unknown and difficult-to-assess conditions than in new building construction. Among other things, these may cause the indoor climate in the completed building to be of lower standard, and manifest themselves as undesirable odours and airborne pollutants of varying degrees of hazard to health. The extent of the problem varies from levels of pollutants so low that they are not noted at all, to levels where most of them expose a building's occupants to serious and immediate symptoms. Such defects are often expensive to remedy. Assessment of the status of the existing building, with its subsequent demands for control, planning and implementation, also exert powerful economic pressure. There is, however, a shortage of established tools and procedures, and a lack of fundamental knowledge regarding how to examine factors that would secure a healthy indoor climate before refurbishing or rebuilding.

The main objective of the research described in this chapter is to develop methods for examining and assessing parts of existing buildings. A secondary objective is to develop the methods to evaluate the future performance of the affected parts in the renovated building. Another objective is to develop methods and tools to diagnose more confidently the damage and indoor air problems in buildings so that it is possible to choose adequate, reliable and cost-effective remedial measures. The final objective is to transmit the results from these assessments to the parties that are involved in the refurbishment or rebuilding project.

State-of-the-art review

Problems faced in buildings

The term 'unhealthy building' refers to a building in which people feel unwell because of the building. The symptoms associated with this are summed up under the recognised term 'sick building syndrome', or SBS. A better term for explaining this condition is 'building related illness', or BRI. This interdisciplinary problem area has been amply described in scientific literature and interests researchers from large parts of the world, but mainly those in the industrialised countries (Samet 1993).

The greatest cause of BRI is the presence of chemical and/or biological airborne pollutants in the indoor environment, in concentrations that give rise to ill health in certain people. Examples of such pollutants are volatile organic compounds (VOC) and particles (Koponen *et al.* 2001). Pollutants in the indoor environment may originate from outside, but also from the inside through emissions from, for example, building materials, fittings and furnishings, building services, people themselves or the activities in the building.

The term 'emissions' is used here as a name for both chemicals and biological emissions of gaseous substances, particles and fibres, and for radiation. In the case of materials, pollutants stem mostly from primary emissions. These occur without any damage to the material. More significant pollutants arise through secondary emission from materials after they have been damaged, altered or broken down. Examples of these are emissions from the adhesive on plastic floor coverings that has decomposed due to alkaline moisture from concrete (Sjöberg 2001), and high moisture levels in organic materials that give rise to abnormal growth of micro-organisms.

The causality between ill health and measurable pollutants in the indoor environment is largely unclear. However, it has often been found that the sources of pollutants that are the cause of BRI are emissions from surface finishes and building components (Gustafsson 1992; Walinder *et al.* 2001). A practicable way of remedying defects in buildings with BRI may be to look for abnormal primary emissions or secondary emissions from damaged finishes and building construction (Wessén & Hall 1999). Assessment should, however, be made with care and be preceded by a broad-based and well-executed damage investigation.

The experience of indoor air quality specialists shows that the causes of pollution in indoor air can be found and traced, for instance, to moisture-damaged building materials or construction. When these damaged building components are subject to remedial treatment, the symptoms decrease or vanish (Kumlin *et al.* 1994; Samuelsson *et al.* 1999).

This knowledge is, together with countless investigations of the cause of BRI and the author's experience over 15 years as an indoor air quality specialist, the foundation of and, to a considerable extent, the basis of subsequent sections. This knowledge and experience are utilised as a basis for the research, in order to enhance the quality of proposed remedial measures and their implementation in problematic buildings, by providing it with a better scientific foundation.

The need to verify or assess the causes and sources of BRI by technical measurement does not arise only when a building is judged unhealthy. Measurement is also needed after new construction, refurbishment or rebuilding in order to verify that the building has a healthy indoor climate, or at least to ensure that the finishes and construction do not have elevated levels of emissions. An appraisal of air quality and any abnormal emission sources may also be desirable in existing buildings because an owner or a tenant, for a number of reasons, requires it.

Appraisal of, for example, moisture status, emission levels and any pollutants that may have been deposited from previous activities or high-emitting materials may also be necessary prior to a rebuilding project. This should provide knowledge of how existing materials, building construction and services installations must be remedied in order that the building should have a healthy indoor climate. The same applies regarding the care that must be exercised when existing and new materials are combined or when composite materials are used.

In the case of rebuilding, there are often a greater number of unknown and difficult-to-assess conditions than in new construction, such as the conversion of an industrial building to offices. For instance, there may be unknown causes of damage in materials and building construction, with associated elevated emissions of a long-term character caused by previous high moisture levels. There may be deposits of pollutants, long inside existing structural elements, from both previous activities and from older high-emitting building materials. There may exist materials and types of building construction, which, in combination with modern materials, may create new problems in the altered building. These may result in an indoor climate of a lower standard in the completed building, and may manifest as undesirable smells and in various degrees of pollutants that are hazardous to health; for instance, high levels of xenobiotic compounds.

In appraising the condition of buildings that are not known to cause BRI, the approach of looking for damaged materials or other defective functions etc. may also be employed. It should, in this way, be possible to assess empirically whether the building has normal or abnormal emissions from materials and building construction. In the same way, it should be possible to judge the concentration of pollutants in the indoor air itself. A moisture status analysis can also be performed to assess the risk that damage may have occurred or is likely to occur in future; for instance, the risk of microbial growth or an elevated content of chemical emissions.

With a properly executed damage investigation, it is often possible to find the cause of defects in buildings with BRI. On the other hand, it is not possible, using the same knowledge and tools, to decide by technical measurement that a building cannot in the future give rise to BRI. The reason for this is that there is very little coupling between health effects and the content of different pollutants. It is, for instance, impossible with the present state of knowledge to assess by technical measurement alone whether a new or altered building is free from pollutants hazardous to health. Instead, the occupants themselves must after completion act as the decisive measuring instruments.

We can only check on those pollutants for which measurement and analysis equipment is available. Based on these results, we can empirically judge whether these levels correspond to what we usually measure in healthy buildings or in buildings with known BRI. Obviously, this is highly unsatisfactory. In this case, also, the solution may be to develop knowledge by in-depth research so that materials and material combinations and their effects on the indoor environment may be better assessed.

Future emissions

One of the more important parameters regarding the physical status of an existing building is that of future emissions.

The future emissions from the materials used in a renovated or new building will depend on the properties and composition of the materials, including residual substances that may be emitted, and the future environmental conditions inside and around those materials. Materials that do not, or did not, give off any emissions in the existing building may be significant sources of emission in the future if the environmental conditions change or if the material is combined with others in a new way.

Methods for examining the materials used, including possible contaminants, with respect to critical transport properties, the content of possible future emissions, sensitivity to changed environmental conditions and combination with other materials, have to be studied in order to improve understanding and quantification (Gustafsson 1992).

The rate of emission of substances from the materials in a building is governed by a complex interaction between a number of factors. Certain properties of the materials and substances are critical. A number of liberation and transport processes in the material and between materials in material combinations will be involved – see Fig. 8.1. The temperature and moisture conditions in the materials and at their surfaces will have a significant effect on these processes (Haghighat & De Bellis 1998).

We can identify three cases where the processes in the materials are quite different:

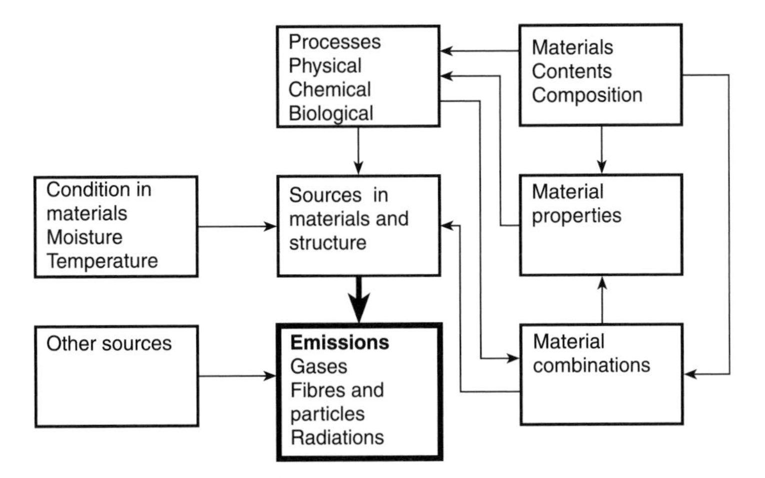

Fig. 8.1 Emission is governed by a complex interaction between a number of factors.

- primary emissions from a single material;
- secondary emissions of future reaction products from material combinations; and
- secondary emissions of absorbed substances.

Primary emissions from a single material

Some substances from building materials, and other materials within a building, may be emitted as a gas into the indoor air, if the substance is not firmly bound and exceeds its equilibrium content in the material. Such an emission process usually involves a process of liberation of a portion of the substance from the material and a transport process up to the material surface or a vaporisation front, where the substance turns into an emission from the material. There are numerous typical examples; for instance, rapid emissions from thin surface finishes or surface materials on floors, walls and ceilings, long-term emissions from thicker building components etc. (Gustafsson 1992).

To be able to describe and predict the rates of primary emission, the emission process has to be understood and quantified. The binding and transport properties of relevant substances, at different levels of temperature and moisture, must be measured. The rate of emission of the substance should be predicted by solving a mass balance equation. The understanding and quantification of the material processes should be verified by comparing predictions with measured distributions of the substance in a material and with measured rates of emission.

Secondary emissions of future reaction products

In certain cases, a substance in a material may react with a substance from another material in combination and give rise to reaction products that may be liberated as emissions from the material combination. The reaction may occur at a certain depth in one of the materials or at the boundary between the materials. Several liberation and transport processes may be involved. A typical example is combinations of vinyl floor coverings on concrete slabs (Wengholt-Johnsson 1995).

The description and quantification of the rate of emission of a reaction product must be based on an understanding of the material processes involved and on measurements of the critical properties of the materials and the relevant substances.

Secondary emissions of absorbed substances

In an existing building, materials may contain substances absorbed from the indoor air or from the previous activities in the building. Additionally, substances from other materials or reaction products from earlier deterioration, for example from floor coverings exposed to moisture and alkali attack, may have penetrated deep into the substrate materials (Sjöberg 2001). Methods for sampling or measuring the distribution of such substances in situ must be developed and prediction methods for future behaviour must be developed and validated.

Physical status of an existing building

Other important factors and parameters regarding the physical status of an existing building are diagnosing damage and problems, residual service life, moisture mechanical behaviour, material properties and environmental assessment.

Diagnosing damage and problems

To make a diagnosis is to decide the cause of damage and problems that have occurred. A diagnosis is built on data concerning the damage history and findings from an investigation. By analysing this information, the cause is determined in such a way that it explains relevant findings and is shown to be able to give rise to the damage history in question. If this cannot be done completely, the diagnosis has a certain degree of uncertainty. If the result of the analysis indicates that several causes may be possible, additional information or investigation is needed to distinguish between them.

To provide the basis for an opinion regarding liability, it is usually necessary to carry the diagnosis so far that it is possible to show that causality exists, i.e. there is a connection between the precise cause of the damage and someone's responsibility. On the other hand, if the purpose of the diagnosis is only to provide the basis for repair measures, the diagnosis does not have to distinguish between different causal connections that require the same measures. The same general approach can be applied to most types of damage. However, this section concentrates on moisture-related damage and indoor air problems.

Moisture may cause damage to building materials, give rise to unhealthy emission to the indoor air, and increase allergic symptoms. If adequate measures are taken in a damaged building and its structure, emissions and, thus, allergic symptoms will be reduced (Ekstrand-Tobin & Lindgren 2000). Damage due to moisture and indoor air problems in buildings is a frequent occurrence (Walinder *et al.* 2001). The cause of the damage and problems must be known in great detail so that the proper remedial measures may be recommended and liability correctly determined. However, such a diagnosis is almost never based on scientific grounds but is more or less an informed guess. A diagnosis must satisfy some requirements if it is to be regarded as reliable (Nilsson 1994). These requirements may be expressed in this way for moisture damage. A moisture diagnosis must show:

- a connection between moisture and damage, i.e. show that the damage is due to moisture and not something quite different;
- the origin of the moisture, i.e. from where the moisture was supplied; and
- the causality, i.e. why this moisture source could cause the damage.

The causality is shown if these requirements are satisfied, and a cause is found if it explains all the information and findings and leaves no loose ends. If this foundation does not exist as a basis for the conclusion, the diagnosis is not reliable.

Residual service life

The service life of building materials and components depends on the environmental actions, critical material properties and the functional requirements for those elements. A prediction of the service life involves describing performance over time in terms that can be compared with the functional requirements (Aktan *et al.* 1997).

When evaluating the status of an existing building, the first important step is to estimate the remaining service life of the structural elements, the materials that are critical for the structural integrity of the building and those parts of the piped and ducted systems that cannot be replaced without demolishing large parts of the building. The next important step is to evaluate the residual service life of non-structural materials, i.e. exterior and interior surface materials, insulation etc.

The residual service life is a function of the previous and future environments around the materials and components. The environmental conditions in the existing building may be accurately determined by direct measurement or estimated from the response of selected materials, by measuring moisture distributions and depths of penetration of certain substances etc. (Sjöberg 2001). The critical transport properties of the materials can be measured directly, in situ or from samples. The predictive models for performance over time can be extensively calibrated on one occasion by examining the present values of the performance parameters in the existing building. In this way, predictive models for describing the future behaviour of the materials and components in the same building after renovation, and possibly altered environmental conditions for the materials, may be significantly improved. Even the predictive models for future utilisation of the materials and components used in a new building may be improved largely if the knowledge of the earlier behaviour in a somewhat different environment is taken into account.

Tools, methods and models for evaluating environmental parameters and critical material properties from materials and components in an existing building have to be improved, based on a better quantitative understanding of the deterioration mechanisms for various materials and components. Some areas of research and some materials in certain applications are *mature*. Other areas are *immature* and require thorough studies at great depth and over long periods. There is little basic knowledge and experience is lacking. The time taken to develop new applications is long and much support is needed over several years.

Moisture mechanical behaviour

The design of a building or a building component is undertaken, among other things, to avoid moisture damage, moisture-induced emissions or moisture-induced changes in the performance of materials. One result of *moisture design* is that changes over time in moisture distributions within materials and components can be predicted (Nilsson 1984). Consequently, in an existing building, those results are of critical importance for predicting the residual service life of materials and components as well as for predicting future rates of emission from the materials used.

Moisture design requires knowledge of the critical moisture properties of the materials, transport properties, moisture fixation properties and critical limits, as well as knowledge of future environmental actions on the materials and components. With proper calculation procedures, moisture mechanical behaviour can then be predicted and used in design.

In an existing building, the composition and the quality of the materials may not be well known. However, the properties and moisture prediction models may be derived and calibrated by taking measurements on the existing structures and comparing them with the predictions. The true moisture mechanical behaviour of existing materials and components in a given environment would make it possible to improve significantly the prediction models for both new and existing buildings.

Material properties in situ

Used materials may, in some respects, have the same or better properties than new materials. Before a demolition process is planned, selected materials should be examined to estimate their properties for a variety of future uses in new buildings. Such estimates may be derived from in situ measurement, partly destructive and partly non-destructive (Wessén *et al.* 1999; Wolkoff *et al.* 1991) or from laboratory or field measurements on statistically selected samples.

A study is needed to determine which are the properties that are most significant in making estimates about a selection of materials, before a decision is taken on how to extract them from an existing building – acceptance levels for evaluating the test results must be expressed. Methods for performing measurement have to be developed or adjusted to the special conditions for used materials.

Environmental assessment

The assessment of the status of an existing building before decisions are made on maintenance, repair, refurbishment or rebuilding must include an assessment of the environmental aspects of the building materials and components in the building. This topic is extensively covered in several recent and on-going projects, i.e. within the Research School of Sustainable Building (Glaumann 1996), supported by MISTRA (The Foundation for Strategic Environmental Research).

Research project

The sub-theme of future emissions has been selected for the research project covered by the chapter. A major reason for this selection is its significance to the future use of an existing building that is to be refurbished or rebuilt, possibly for quite new occupant activities. Other reasons are the present lack of knowledge and experience in this area and the urgent need for tools to quantify the present status and predict future behaviour.

Description and objectives

The study is limited to developing tools to control future emissions from contaminants in building materials in existing buildings. The objectives are to develop methods to:

- survey the distribution and measure the penetration profiles of absorbed substances;
- predict future emissions of substances from materials and structures; and
- assess remedial measures against future emissions.

Research methodology

The methodology comprises five stages: first, measuring the content and distribution of substances that can be emitted in materials and structures; second, modelling the rates of emission based on an understanding of the chemical and physical processes; third, quantifying the critical properties of the materials and the substances in independent measurement; fourth, verifying the validity of the models in controlled laboratory experiments and in field studies of rates of emission from single materials and from combinations of materials; fifth, guidelines for practical applications, for example, on the choice of material combinations, testing of materials and design of structures, based on the models developed in the programme area.

Project plan

The project begins with a broad approach. Existing knowledge is identified, elicited and structured. Laboratory experiments will be performed in order to create penetration profiles of *absorbed substances* under controlled conditions. Field experiments will be performed in several ways. A preliminary aim is that 5–10 buildings will be selected for study. Some should be suspected of having contaminants from the previous activity in them. Some should be suspected of having traces of reaction products from, for example, flooring materials.

Attempts will be made to detect the presence of substances from the materials in the indoor air by analysing air samples. Rates of emission will be measured at selected surfaces of the original materials and at various depths inside the materials.

Air samples for these measurements will taken with a Field and Laboratory Emission Cell (FLEC) (Wolkoff *et al.* 1991). Methods for surveying the distribution of substances in the structures will be tested and possible development identified and undertaken. Other methods will be tested to measure the penetration profiles of substances at statistically selected points in the building's structure and at selected points shown by the previous survey to be representative of various elements. Generally accepted methods do not exist today, but two options would be to measure inside small holes drilled into the elements or to take small samples.

Laboratory experiments will be performed before and after the field tests to develop the survey methods, measurement methods and analysis techniques. Simple, inexpensive analysis techniques should be a goal, with affordable thorough surveys and measurements. Laboratory measurements will also be taken on rate-critical parameters of emission processes to be included as input data for the predictive models.

Predictive models for the absorption and emission processes under various conditions and for a selection of materials and substances will then be developed. This is done by quantifying the important parts of the mass-balance equation, from the penetration profiles and from separate measurements on thin samples. The predictive models will be verified by measuring the rates of emission at the materials' surfaces in a few of the selected buildings some years after they were rebuilt. Additional verification will be achieved through laboratory specimens. The predictive models can be used to assess alternative remedial measures. The assessment must also consider the planned future use of the building by incorporating its requirements for indoor air quality.

Research results and industrial impact

The project is expected to yield survey methods and tools for assessing the risk of future emissions in existing buildings that are to be the subject of refurbishment or rebuilding. Such tools and methods are needed by the sector for at least two reasons. First, future building activities will be mostly concerned with rebuilding the existing building stock, which requires a different approach to future emissions. Second, a good indoor air quality is an increasingly important prerequisite for all buildings.

Conclusions

In order to reduce the number of cases of BRI and its consequences in new building, refurbishment and rebuilding, we need knowledge and the ability to understand, and also to predict, how materials and material combinations behave in different service situations. Further research inputs are needed to develop this knowledge as a basis for recommendations as to how materials must be assessed prior to use. This applies to new building and rebuilding, and also to problem buildings under occupation. Such know-how would provide the basis for developing measurement and predictive tools for use in assessments. The cost of a thorough status determination should be compared with the costs of the possible consequences of bad decisions due to lack of relevant information during the redesign of an existing building.

References

Aktan, A. E., Farhey, D. N., Hehmcki, A.J. *et al.* (1997) Structural identification for condition assessment: experimental arts, *Structural Engineering,* **123,** pp. 1674–85.

Ekstrand-Tobin, A. & Lindgren, S. (2000) Indoor environment and allergic parameters before and after renovation of moisture damaged constructions, *Healthy Buildings 2000,* Espoo, **1**, pp. 227–32.

Glaumann, M. (1996) *Environmental assessment of buildings,* Part 1 Background and part 2 Proposals, BFR-report 940583, Stockholm: Swedish Council for Building Research (in Swedish).

Gustafsson, H. (1992) *Building materials identified as major sources for indoor air pollutants,* Stockholm: Swedish Council for Building Research.

Haghighat, F. & De Bellis, L. (1998) Material emission rates: literature review, and the impact of indoor air temperature and relative humidity, *Building and Environment,* **33,** pp. 261–77.

Koponen, I. K., Asmi, A., Keronen, P., Puhto, K. & Kulmala, M. (2001) Indoor air measurement campaign in Helsinki, Finland 1999 – The effect of outdoor air pollution on indoor air, *Atmospheric Environment,* **35,** pp. 1465–77.

Kumlin, A., Åkerlind, L.-O., Hall, T. and Eckerbom, G. (1994) A Method to Solve Indoor Air Quality Problems: A Practical Swedish Strategy, *Indoor Air – An Integrated Approach,* Gold Coast, Australia, **1**, pp. 329–32.

Nilsson, L.-O. (1984) *Moisture problems in renovation of buildings,* Contribution to a Polish-Swedish seminar on Aspects of Renovation and Restoration, Gothenburg: Chalmers University of Technology.

Nilsson, L.-O. (1994) Methods of diagnosing moisture damage in building, *CIB International Symposium on Dealing with Defects in Building,* Varenna, 27–30 September.

Samet, J. M. (1993) Indoor air pollution: A public health perspective, *Indoor Air ´93,* Helsinki, **1**, pp. 3–12.

Samuelsson, I., Fransson, J., Gustavsson, H. *et al.* (1999) *To investigate Indoor Climate and Air Quality,* Technical report, Borås: Swedish National Testing and Research Institute, SP Report 1999:01 (in Swedish with English abstract).

Sjöberg, A. (2001) *Secondary emissions from concrete floors with bonded flooring materials,* Doctoral thesis P-01:2, Gothenburg: Chalmers University of Technology.

Walinder, R., Wieslander, G., Norback, D., Wessen, B. & Venge, P. (2001) Nasal lavage biomarkers: Effects of water damage and microbial growth in an office building, in: *Archives of Environmental Health,* **56,** pp. 30–36.

Wengholt-Johnsson, H. (1995) *Chemical emission from flooring systems – effect of concrete quality and moisture exposures,* Licentiate of technology thesis 95:4. Gothenburg: Chalmers University of Technology (in Swedish).

Wessén, B. & Hall, T. (1999) Directed non-destructive VOC-sampling: a method for source location of indoor pollutants, *Indoor Air 99,* Edinburgh: Scotland, **4**, pp. 420–25.

Wolkoff, P., Clausen, P. A., Nielsen, P. A., Gustafsson, H., Jonsson, B. & Rasmusen, E. (1991) Field and Laboratory Emisson Cell – FLEC, *Healthy Building '91,* Washington DC, pp. 160–5.

Chapter 9

Co-ordination of the Design and Building Process for Optimal Building Performance

Niklas Sörensen

Introduction

During the construction process, many parties are present over a period that can range from months to years. Their interaction can be characterised by the optimisation of numerous factors. Decisions in the early stages of the process usually affect work conducted later. This implies that facts motivating decisions must be communicated throughout the whole process and that decisions should focus on the end product. Failure or lack of communication leads to technical errors, resulting in products that fail to fulfil the client's requirements. The reasons for communication breakdown and subsequent technical failures must be investigated to improve the end product, namely a building that functions as intended.

An objective in the construction process should be a product that works properly in a technical sense and that satisfies the needs of the client. This can be achieved by improving the interface between building technologies, actors who control the different technologies and by having common objectives during the process. The reduction of errors that would occur from a co-ordinated interaction and common objectives could lead to a significant decrease in capital cost for the project and subsequent operational costs.

One area considered in this chapter is building physics. The subject includes transfer of energy, moisture and air in buildings. High-energy consumption may be indicative of a malfunctioning indoor climate if it is at odds with the expected range of performance. An accurate calculation of energy use is dependent upon well-defined design criteria, design practice, production and installation of services. Lack of attention to building physics is considered to be one of the factors leading to sick building syndrome.

State-of-the-art review

Areas for enhancing building performance

The construction process is exposed to increased complexity (Formoso *et al.* 1998), suggesting that the process itself needs to be systematically reviewed to generate higher value. Interest has been directed to the production philosophies of Japanese manufacturing industry to highlight just-in-time delivery, quality and, in some cases, procedures to simplify process improvement. Interest has, however, grown rather slowly (Koskela 1992).

Attention seems to have been drawn to a limited number of elements of Japanese manufacturing philosophies, where some are introduced to the management role in the construction process; while other parts and preconditions have been little discussed. Reduction of project duration is one issue that draws attention. Wilkinson (2001) found that the main reason for using construction management or project management company services is so that the client achieves speedy completion of the project. It has also been argued that the project management role, in the UK in particular, places too much emphasis on the achievement of time, cost and functional targets that are frequently poorly defined (Brown *et al.* 2001). Ballard (2000) recognised that the theory and techniques of project management are mainly devoted to the formation, award and administration of contracts. This would probably be true for the construction process in many countries.

One issue that seems to be little discussed is that to be able to have an effective procedure, an objective has to be present. This is a growing issue since new building codes are moving away from prescriptive technological codes; e.g. the thermal transmittance for a part of the building envelope must be less than a specified value and conform to a functional code. Established practice is based on such prescriptive requirements, which are perhaps simpler to work with than performance requirements, but they risk standing in the way of more efficient and economical solutions. The reasons for this are that such practice excludes the possibility of another solution being better and that it is a poor way to utilise other people's ideas or knowledge (CIB 1982). With this in mind, the process should start with a thorough investigation focused on the end product to reveal the necessary criteria, allowing them to be communicated through the whole process to the operational phase. If this were to be the end of the process an important phase would be excluded; that is, the evaluation part, where one determines how well the product satisfies the criteria stated in the process. This implies that the criteria should also be quantifiable in some manner.

Establishing project objectives and corresponding technical criteria

A construction project is induced by client needs; but often the project outcome fails to satisfy them. There are many reasons for this and one challenge is to comprehend client needs, which should be revealed during the briefing stage of the project. According to Barrett *et al.* (1999), briefing is seen as a singular event at the beginning of a project, rather than a process, where requirements are systematically written down. This means that while the project proceeds and client awareness increases, the ability to make changes is reduced. The recommendation is that briefing is a process running throughout the construction project, by which means the client's requirements are progressively captured and translated into effect (Barrett *et al.* 1999). It should also be recognised that the changes in design objectives or criteria – some of them introduced by clients – are the kinds of factors that tend to push the design process away from the optimal track (Koskela *et al.* 1997). However, this does not mean that the process will be straightforward.

Construction clients are buying something they cannot see until it is completed. Few clients can read drawings in a meaningful way and effective communication is therefore needed that highlights coded information, a robust message that is reinforced and fed back (Barrett *et al.* 1999). Barrett (2001) found that failing to understand client needs is the issue that creates the largest gap between client expectation and client satisfaction.

The second largest gap is held by project delivery on time. The briefing should, therefore, not only articulate the strategic objectives, and fit into wider corporate, risk and environmental management strategies, it should also move assumptions out into the open so that they are not wrongly second-guessed by the designer (Bordass & Leaman 1997). This is a problem that the client has to consider, since there are likely to be different and contradictory needs within his own organisation. It is also important to make the distinction between *paying client* and *end user* (Barrett *et al.* 1999).

During the design process, one issue that is raised is the lack of information to complete tasks (Koskela *et al.* 1997). However, it is also true that the designer needs to break down design tasks in order to identify what needs to be done to make assignments ready to be performed (Ballard 1999). With this in mind, it seems as if the information that is needed in design is not only dependent on the transmitter of the information, but also on the receiver.

One challenge is how to translate client needs into design requirements and subsequent critical characteristics. Gargione (1999) has used the Quality Function Deployment method (QFD), developed by Akao in 1990, for this purpose. QFD is a method that a) develops design quality aimed at satisfying the consumer and b) translates the consumer's demand into design targets and major quality assurance points to be used throughout the production stage. Gargione identified some obstacles in using QFD; for instance the focus group did not express most of their opinions, needs and requirements clearly enough. This could be based on the possibility that participants might be unwilling to share knowledge even if they were brought together face-to-face, in the hope that the final solution better favours them than others (Ballard 2000). This leads to results originating from discussions and negotiations within the client body and between professionals with diverse perspectives (Kamara & Anumba 2001). It also transpired that the time to plan and analyse data increased substantially.

Despite the problems, a method to convert client requirements into design criteria is useful even though prioritising and keeping data up-to-date is still a major challenge. One question is: what should the project objectives be? Brown *et al.* (2001) emphasise that the long-term operation of the end product should be a primary objective of the process and that one of the reasons for this matter to fail is that the short-term business needs of those contributing to the project interfere with the project's efficient management. This seems to be relevant, but there are also other forces driving the contributors; for instance, cognitive aspects of motivation and commitment (Bresnen & Marshall 2000). If the long-term operational phase of a building is the main priority of a project, then who is most able to define this requirement and convert it to design and production requirements?

Brown *et al.* (2001) believe that the client is the party that benefits from the long-term operation of the building and should therefore lead the process, but it is also clear that the client is not interested in technological correctness (Barrett 2001). The designer on the other hand has the technological competence, but handling all the interdependencies to reach an optimal technological solution can sometimes lead to long design and project duration (Ballard 1999). The designer also has less knowledge of how to produce a product that is designed (Alarcón & Mardones 1998). This production knowledge is the speciality of the contractor and subcontractors, but sometimes they rely too heavily on the fact that it should be easy to produce. Apparently no single party is fully capable of leading, but rather a group of individuals can stand a better chance of succeeding.

A number of key issues can be observed:

- Clients should have enough time to increase their knowledge of the project outcome, based on their requirements.
- Clients should be able to change their mind when the challenges of their requests are made apparent to them.
- Designers should have sufficient time to convert client requests into key technical criteria.
- Designers should have enough time to investigate the interdependencies of the technical criteria in the building system.

- Contractors and subcontractors, when required, should have the ability to view the impact of decisions regarding constructability.

This calls for a better understanding of the criteria for project success before commencing the subsequent phases of a project.

Supervising the process

One issue is how design criteria that support client needs and form a well-performing building, when successfully stated, should be handled through successive stages of the process. Even though there is a phase that systematically converts client requests from the brief into design criteria, that also takes the construction stage into consideration, it is inevitable that issues relating to project success will have been anticipated. For instance, clients must be allowed to change their minds, designers will find interdependencies that were not foreseen and during the production phase there will be new conditions that were previously unknown.

As earlier identified the client comprehends the product increasingly as the process proceeds. This is not only true for the early stages of design, but also during detailed design and the production stage. This suggests that even though it is hard to manage, it should be possible to review the requirements of the client in order to produce a building that satisfies. Barrett *et al.* (1999) state that when clients attempt to change the parameters, as their knowledge and feel for the issues involved increases, they are invariably reminded of their original statements as if to imply that these cannot be changed. This is perhaps a little exaggerated, but handling changes is increasingly hard to manage as the project progresses. Since the number of interdependencies grows significantly with time, this is one of the reasons for the unwillingness to implement changes. However, it is not only the client who instigates changes; others may feel that they are also necessary. In the design and construction stage, details that imply changes might surface. In such cases, the client ought to be informed and allowed to review the change.

The design process is complex and has to handle a number of interdependencies. The process also oscillates between criteria and alternatives (Ballard 2000). These facts imply that it is hard to plan and manage the process of design, but it is still necessary. An area that lacks support is the integral comparison of different design alternatives (Augenbroe & Pearce 1998) that would help in understanding the impact of changes. Any changes should be verified in order to satisfy the main objectives of the project stated earlier in the process.

The building construction stage should be easy to plan and manage, if:

- the client is satisfied with the design;
- the design is correct and can be realised through construction;
- the intention of the designer is correctly communicated; and
- all conditions on site can be anticipated.

This is seldom the case, but if the pre-existing phases were improved the building construction phase could be much easier to manage.

In order to manage the process, a regular review against strategic objectives must be present (Bordass & Leaman 1997). If they are not, the briefing becomes the set of papers on the shelf that Barrett *et al.* (1999) recognised. When changes are necessary, the interdependent areas of responsibility must be informed and suitable actions take place. Working with monitoring changes will allow a meaningful feedback to all actors revealing what was achieved, what was not achieved, why it was not achieved and what can be done to improve the project's outcome.

Evaluating project results

It seems as though projects have a tendency to end with the statement that 'we made it possible despite all the problems we had with the other participants', without acknowledging the obstacles encountered and the improvements effected. An example where this would be insufficient is in the house-building sector where evaluation of energy efficiency is a crucial area. Good evaluation helps break through the confidence barrier by countering owners' and occupiers' scepticism about energy savings and the level of the investment required. The evaluation of findings is also essential for helping researchers and practitioners to refine their methods (DeCicco *et al.* 1994). Measuring the outcome is therefore important and it can only be performed if there is something with which to compare it. Therefore, the earlier briefing process should reveal quantifiable parameters rather than subjective ones, i.e. a building that 'I believe fulfils my requirements'. It is also important that all actors involved in a project are satisfied with the outcome.

Research project

A performance approach is foremost the practice of thinking and working in terms of ends rather than means (CIB 1982). In order to improve overall performance, the construction process has to increase its focus on building technologies and their interdependencies in the building system on a long-term basis. It is important to investigate criteria thoroughly in the early stages, because of the evolving nature of the process. This must occur before releasing subsequent phases in order to maintain focus on these criteria through the project.

Role of building physics

Building physics is an area that has been somewhat neglected, yet it is a subject that is closely connected to building performance. It involves many building technologies that should receive systematic attention in the design phase, and is also dependent upon customer requirements, performance on site and commitment in use. The increased complexity and number of specialists involved in the process call for better communication and co-ordination of issues involving building physics.

Neglecting issues involving building physics has been shown to lead to failures in terms of: high energy consumption, moisture problems, indoor climate problems and sick building syndrome, among other conditions. In order to improve the conditions for project success, in the context of building physics, a number of issues have to be investigated.

- What goes wrong? Findings have to be prioritised to streamline the management of the project, since there is a substantial amount of information to consider.
- What is the reason for failure? Is it due to poor communication, lack of objectives, wrong technologies, errors in installation etc?
- What can be done to reduce the incidence of failures in projects?

These investigations have to include the client briefing, design, construction and operational phases of the process. The design stage is the primary target, since it is where the product is defined. The results could lead to new forms of project organisation, new objectives in projects, new roles in projects and new forms of procurement.

Research methodology

The issues addressed in improving building performance relate to the behaviour of actors in the construction process rather than to the technologies that are available. A substantial number of problems have been observed, but many seem to be repeated and so the reasons for this should be investigated.

The study is being conducted as a form of applied research with theoretical underpinnings (Jensen 1991). This theoretical foundation is based on research within building physics, communication, organisation and process. The research will contain a quantitative part, i.e. where errors in the performance of building technology and systems are measured, and a qualitative part, i.e. where reasons for failures are drawn from the actors involved. The aim is to understand why failures arise.

The focus group will be clients, consultants and contractors, because of their close interaction and the nature of the challenges involved. This will also give a natural triangulation, i.e. all parties are included and their opinion is considered, which should improve the generalisation of the results.

The research questions and hypothesis are based on empirical regularity (Alvesson & Skoldberg 1994), something that has been recognised by consultants and other actors in the construction sector. There is, however, a risk of subjective contribution to the questions and hypothesis, but if the research is correctly conducted this will be without prejudice to the objectivity and applicability of the results.

The research will oscillate between theory and empirical studies, which leads to a risk that research questions and hypothesis will be adjusted to fit new discoveries. This suggests that during the progress of the project the research might deviate from the subject (Jensen 1991). This is countered by a research objective that is continuously monitored and when changes are made they can be traced. Results from the research will be implemented in building projects allowing confirmatory applicability.

Project success from concentrating on building performance

It is suggested that improved building performance is connected to a better understanding of building physics, that would lead to lower energy consumption, better indoor climate, fewer defects due to moisture and a decrease in the incidence of sick building syndrome.

Establishing objectives, co-ordinating and measuring long-term criteria

Energy consumption is an area where long-term operation has to be taken into account and strategic criteria for energy-efficient buildings have to be established. In Sweden, the energy use calculation, where required energy for heating is determined, is seen as a singular event at the beginning of a project where the calculated energy use for the building must be less than the energy use stated in the building codes. Calculations are usually based on experience and expectations regarding insulation, window quality, air change ratio, heat recovery etc., with little involvement of the client. Examining energy issues with input based on experience alone is not sufficient, since correct input is vital for relevant output (Johannesson 1993). The tools used in such calculation work are also not applied in a systematic way and may not, therefore, establish interdependencies and allocate potential reductions and efficiency in energy use.

As noted above, energy calculations are often seen as a self-contained task and thereafter abandoned. This is seriously misleading, since many factors interdependent with energy use become evident later in the process, and changes are made to factors established earlier. For instance, it could be assumed early that the thermal transmittance value for windows would be at a certain level, but this could be changed both during design and construction with a substantial influence on the average thermal transmittance for the whole building. Since many changes are made, it is not only important to supervise the process but also to measure the outcome: in this case the actual energy consumption. Working this way reveals how well the building corresponds to the levels of consumption previously calculated. When large differences are discovered there is the possibility of allocating the criteria involved and taking suitable action. This information also serves as a foundation leading to improved work with energy issues that will decrease long-term operational costs for clients, as well as reducing the overall environmental impact of the sector.

Indoor climate is an issue that is vulnerable to subjective understanding that challenges engineers and other actors in the construction process to understand and articulate client requirements. It is of great importance to state objectives and convert these into technological factors, allowing clients to receive a building that satisfies their needs. However, some desires cannot be satisfied and others are expensive to achieve; sometimes they cost more than the client expects. An example is indoor temperature, where a small variation between minimum and maximum temperature leads to a greater energy use than when a large variation is allowed. Many clients would probably accept a larger variation in temperature when its relation to energy use is explained. Moisture in buildings leads to the fabric decomposing at a greater rate than expected and therefore increases maintenance costs, as well as leading to dissatisfied clients. These issues are tied closely to the increased complexity of buildings and the use of new materials and can be substantially prevented by the systematic investigation of materials and their suitability for incorporation in the building.

These examples illustrate the need for a systematic review of the building system with relevant objectives. Improved communication and close monitoring of the process will lead to better technological solutions and fewer failures in the area of building physics.

Possible application and gains

If an increased effort could be put into stating objectives, and understanding building technologies and their interdependencies in the early stages of a project, an increased level of efficiency would be achieved. When a building is viewed as an integrated system rather than a series of independent components, it can have a significant effect on the outcome (Augenbroe & Pearce 1998).

In a survey of seven Swedish construction projects it was found that 4.5% of production costs were due to failures. A quarter of these failures were attributed to design, and the main cause was lack of co-ordination, leading to conflicting actions (Josephson & Hammarlund 1997). A close monitoring of the process, that allows for action when needed and following up on the end product, would also improve the predictability of projects and reveal weak points. An example of how changes could affect the project outcome is energy consumption for housing exceeding calculated levels by 50% (Adalberth 1997). Some reduction would be possible by closer monitoring of the process, allowing it to approach the calculated values. This is a major challenge for the sector that might lead to changes in project life stages, project organisation and project actors. With an increasing complexity and greater demands on achieving expected project results an improvement in these areas is likely.

Conclusions

The task of converting objectives to technological criteria is essential to achieving project success. This is largely self-evident, but implies thorough investigation of interdependent factors throughout the project so that expectations and likely end results can be compared. In this way, serious and potentially damaging deviations can be avoided.

This raises the question of the extent of technological errors that might be attributable to the failure to establish project objectives before work is undertaken, and the reasons for their occurrence. It is also important to understand the reasons for technological failures due to communication breakdown in the briefing process, and during subsequent life stages of a project. Another question is how measuring project results can be used to improve the sector's performance and the factors that should be made quantifiable. Improvement in these areas within building physics will improve energy conservation, indoor climate and reduce moisture problems with a potential decrease in the number of sick buildings as a result. The answers to these questions rest with the design stage of a project, but are also dependent upon all actors involved across the different life stages.

References

Adalberth, K. (1997) Energy use during the life cycle of single unit dwellings: examples, *Building and Environment*, **32**, *4*, pp. 321–29.

Alarcón, L.F. & Mardones, D.A. (1998) Improving the Design-Construction Interface, *6th Annual Conference of the International Group of Lean Construction*, Guarojá, Brazil, 13–15 August.

Alvesson, M. & Skoldberg, K. (1994) *Tolkning och reflektion: Vetenskapsfilosofi och kvalitativ metod*, Lund: Lund Institute of Technology (in Swedish).

Augenbroe, G. & Pearce, A.R. (1998) *Sustainable Construction in the United States of America*, CIB-W82 report, Atlanta, GA: Georgia Institute of Technology.

Ballard, G. (1999) Can pull techniques be used in design? *Conference on Concurrent Engineering in Construction*, Espoo: August.

Ballard, G. (2000) Managing work flow on design projects, *CIB W96 Conference*, Atlanta, GA: 19–20 May.

Barrett, P., Sexton, M. & Stanley, C. (1999) Key improvement areas for better briefing, *Proceedings of Joint Triennial Symposium (CIB Commissions W55 & W56 with participation of W92), Customer Satisfaction: A Focus for Research & Practice*, Cape Town, South Africa, 5–10 September, **1**, pp. 535–45.

Barrett, P. (2001) A survey of construction clients' needs, *CIB World Building Congress 2001*, Wellington, New Zealand, 2–6 April.

Bordass, B. & Leaman, A. (1997) Building services in use: some lessons for briefing, design and management, *BIFM annual conference 1997*, London, 17 September.

Bresnen, M. & Marshall, N. (2000) Motivation, commitment and the use of incentives in partnership and alliances, *Construction Management and Economics*, **18**, pp. 587–98.

Brown, A., Hinks, J. & Sneddon, J. (2001) The facilities management role in new building procurement, *Facilities*, **19**, *3/4*, pp. 119–30.

CIB (1982) *Working with the Performance Approach in Building*, Publication 64, Rotterdam: CIB (International Council for Building Research Studies and Documentation).

DeCicco, J., Smith, L., Diamond, R., *et al.* (1994) Energy conservation in multifamily housing: review and recommendations for retrofit programs, *Proceedings of the ACCE 1994 Summer Study on Energy Efficiency in Buildings*, **10**, Berkeley, CA: 18 August–23 September, pp. 10.21–10.33.

Formoso, C.T., Tzotzopoulos, P., Jobim, M.S.S. & Liedtke, R. (1998) Developing a Protocol for Managing the Process in the Building Industry, *6th Annual Conference of the International Group of Lean Construction*, Guarojá, Brazil, 13–15 August.

Gargione, L.A. (1999) Using quality function deployment (QFD) in the design phase of an apartment construction project, *Proceedings of the 7th Annual Conference of the International Group of Lean Construction*, Berkeley, CA: 26–28 July, pp. 357–368.

Jensen, M.K. (1991) *Kvalitativa metoders anvandning i samhallsforskning* Copenhagen: Socialforskningsinstitutet (in Danish).

Johannesson, C.M. (1993) *Stockholmsprojektet-utokad varmeisolering och kvalitetsstyrning*, Report R12:1993, Stockholm: Byggforskningsradet (in Swedish).

Josephson, P.-E. & Hammarlund, Y. (1997) The cost of quality defects in the 90s: There is a lack of commitment, *Swedish Building Research*, **4**, pp. 2–4.

Kamara, J.M. & Anumba, C.J. (2001) A critical appraisal of the briefing process in construction, *Construction Research*, **2**, pp. 13–24.
Koskela, L. (1992) *Application of the New Production Philosophy to Construction*, Technical Report No. 72, Stanford: CIFE, Stanford University.
Koskela, L., Ballard, G. & Tanhuanpaa, V.-P. (1997) Towards lean design management, *Proceedings of the 5th Annual Conference of the International Group of Lean Construction*, Gold Coast, Australia, pp. 1–12.
Wilkinson, S. (2001) An analysis of the problems faced by project management companies managing construction projects, *Engineering, Construction and Architectural Management*, **8**, 3, pp. 160–70.

Chapter 10

New Concrete Materials Technology for Competitive Construction

Markus Peterson

Introduction

Cast in situ concrete is often passed over in house building in favour of other materials such as precast concrete and structural steel. The justification for selecting other materials is generally due to their perceived shorter production times and longer spanning capability. Novel concrete materials technology can, however, counter these disadvantages. Extensive international materials research into new concrete technology, especially high-performance and self-compacting concrete, has revealed new opportunities for the design, production and function of concrete construction in low/medium-rise house building. The implementation of new materials technology has occurred in civil engineering work and in the construction of high-rise buildings. In low/medium-rise house building, cast in situ concrete is used almost in the same way and by the same kinds of organisation as in decades past, despite the increasing competitiveness of other materials technology.

The aim of the research reported here is to investigate the potential for new cast in situ concrete materials technology as a competitive alternative for the construction of structural frames. The focus is on design, production and function-related aspects. Economic and organisational aspects are also considered, because of the implications for practice.

State-of-the-art review

Concrete materials research is performed extensively around the world. During the past decade, considerable effort has been put into high-performance concrete (HPC) and self-compacting concrete (SCC). These new materials (or technologies) have been implemented in cast in situ construction work, especially within civil engineering.

HPC has been used in structures such as bridges, roads, offshore construction and high-rise house building. The incentives for use have been high strength and durability, reduced dimensions and dead load, as well as fast strength development and production cycles. HPC has been used in a number of prestigious high-rise building projects; for instance Petronas Twin-Towers in Kuala Lumpur, and the BFG Building and Japan Centre, *Taurustor*, in Frankfurt (Claesson 1999). In the Swedish house-building sector, HPC has been used to a minor extent to achieve a faster concrete curing and drying process. However, these new materials are seldom used in housing, even if they represent solutions to problems that are attributable to ordinary cast in situ concrete.

Concrete research tends to concentrate mainly on technical aspects. Non-technical aspects concerning, for instance, obstacles to implementation or incentives such as economic benefits, are often limited. Nevertheless, some research has shown that rationalisation is possible when using new concrete technology. In Sweden, research results show practical advantages and cost savings from the use of HPC – see, for example, Hallgren (1993) and Persson (1996) who describe some of the economic benefits. The results tend to typify most research, where just one or two aspects are examined. A total concept that would highlight the range of opportunities available from using this novel technology is lacking. Indeed, the latter point is a fundamental issue for construction process improvement and forms the primary aim of the author's research project.

Cast in situ concrete in house building

Structural frames used in house building are based on a particular material, e.g. cast in situ concrete, prefabricated concrete elements, concrete-steel composites or timber. The market shares of these materials differ around the world. In Sweden, approximately 65% of structural frames constructed in 1998 consist of cast in situ concrete, according to a survey by Mängda (1999). That said, some uncertainty exists, due to the tendency to combine types of material in structural frames. The use of cast in situ concrete in house building is criticised for not being as industrialised or as competitive as other materials and their associated methods of construction. Even though cast in situ concrete frames have been the dominant structural method for house building in Sweden in modern times, it is seen to have distinct disadvantages in terms of design, production and function. Increasing industrialisation within the construction sector means, however, that the use of cast in situ concrete must be developed further in order to survive (Byfors 1999). Before describing the possibility of a more industrialised cast in situ concrete method, by using new concrete materials technology, the present method as adopted in house building will be described.

From a design perspective, cast in situ concrete structural frames for house building are normally produced either with tunnel-forms or in combination with prefabricated floor slabs and/or wall units. The most common structural layout is the slab block with curtain wall façades, that are sometimes prefabricated. The most frequently used type of concrete is an ordinary house-building concrete, defined as a mix with a high water/cement ratio (*w/c* ratio) of approximately 0.60, in which reinforcement is generally non-tensioned. The relatively low structural capacity of this concrete and reinforcement permits floor spans up to approximately 5 m. Cast in situ concrete partition walls normally support the slabs to form a solid cell system. This limits flexibility for the customer and the opportunity for future adaptation. For office buildings, other structural concepts allow greater flexibility and are dominant. Column/slab structures and post-tensioned reinforcement are used to a large extent. These concepts are now being introduced into the house-building sector.

In production terms, traditional low-grade, house-building concrete has a high *w/c* ratio. This type of concrete needs a long drying time before floor coverings can be applied in order to avoid moisture-related problems with adhesives or carpets. Swedish building regulations stipulate the maximum allowable values for relative humidity, measured on the equivalent depth of concrete construction (Boverket 1999). These values depend on the type of material used to cover the concrete floor. For most materials the maximum values are between 85% and 90% relative humidity, which usually requires a drying time for *average* concrete of several months. For acoustical reasons, a thick concrete slab is advantageous (Ljunggren 1995). However, in practice, the thickness of the concrete slab is limited, because of the effect of drying times. Partly for this reason, the highest sound insulation class (in accordance with Swedish building regulations) is seldom reached. Concrete cast in situ is a labour-intensive activity as the material has to be vibrated and compacted, a procedure that may lead to hearing impairment for operatives and the condition known as *white fingers* (Hand Arm Vibration Syndrome or HAVS).

New concrete materials technology

In the 1980s, concrete was developed with silica fume in combination with super plasticisers to give increased strength, thereby creating new possibilities for concrete elements such as columns in high-rise buildings (Walraven 1999). This new material was termed high-strength concrete (HSC). However, the concrete had other properties, such as high durability, and was used in other kinds of construction; for instance, offshore. These new properties led to the name being changed to that of high-performance concrete (HPC). During the 1990s, research and applications of HPC have increased dramatically (Helland 1996). Over the past few years, HPC has been common in offshore construction, bridges and high-rise buildings worldwide. Helland (1999) argues that the concrete sector has changed from a low-tech to a high-tech sector and must continue to develop and implement novel concrete materials technology in order to compete with other materials. According to Walraven (1993), the conservative construction sector has to extend international building codes and get accustomed to the idea of using HPC. Increased knowledge in new materials technology may be necessary for designers, if materials are to be fully exploited and their potential risks are to be properly handled (Walraven 2000). Helland (1996) further argues that international codes must follow technical development if they are to avoid major step changes.

Important properties of HPC are its considerably higher compressive strength and durability than ordinary concrete. High performance is reached by increasing the amount of cement and/or lowering the water content by means of water-reducing admixtures. Other properties include faster strength development, higher tensile strength, appreciably faster drying time, good workability, and a higher grade of resistance against mechanical abrasion and penetration of chemicals. Elfgren *et al.* (1994) describe a major Swedish research programme that resulted in two handbooks on materials performance and the design of HPC (Swedish Building Centre 2000a,b).

High-performance concrete (HPC)

The benefits of HPC lead to opportunities for its utilisation within a range of applications, e.g. housing, bridges, offshore, tunnels and roads. Research on HPC in house building concentrates mainly on high-rise buildings. In the US, HSC was at first used in columns in high-rise buildings to achieve greater height and stiffness and to reduce column sizes (Russel & Fiorato 1994). Hoff (1993) summarises HPC high-rise buildings constructed in the US before 1992. In Asia, for example Japan, the use of HPC is described by Ikeda (1993); in Singapore by Chew (1993); and in China by Chen & Wang (1996). Incentive for using HPC in high-rise buildings in Asia has been its high degree of earthquake resistance (Jinnai *et al.* 1999).

In summarising the use of HPC in high-rise buildings, the incentive in most cases has been the increased strength of concrete columns, allowing greater heights and larger floor areas to be achieved.

The properties of HPC, derived largely from its low *w/c* ratio, provide opportunities in the construction of low/medium-rise houses (Öberg & Peterson 2000). In terms of structural design, the properties of HPC enable significantly larger spans and/or slender construction than is possible with *average* concrete. The drying process for HPC also differs significantly from that of ordinary concrete (Persson 1998). In HPC the water is bound to a greater extent in chemical form, creating a self-drying effect. Furthermore, drying time becomes almost independent of the concrete's thickness, which, for an ordinary concrete, is a critical factor.

From a production perspective, the reduced drying time of HPC allows floor coverings to be applied earlier. This leads to shorter total production times and lower production costs. The benefit of HPC is not only its high final strength, but also its fast strength development, that can reduce production time through earlier formwork removal or post-tensioning of reinforcement. Formwork can be better utilised by having shorter removal cycles. Fast strength development can also reduce problems connected to winter casting, because of the reduced risk of freezing young concrete (Fagerlund *et al.* 1999). Costs for warming and covering the concrete can also be reduced.

The advantages of HPC as a structural design and production solution could lead to advantages also in the function of the building. Larger spans in combination with light, easily dismountable partition walls allow a higher degree of flexibility for the owners and occupants, as well as increased future adaptation and refurbishment opportunities. The fast drying process helps to avoid moisture-related health problems that are sometimes the result of inadequate drying time before applying floor coverings. The self-drying effect can also be used to improve acoustic qualities by allowing thicker slabs to be constructed without extending production time.

Self-compacting concrete (SCC)

Self-compacting concrete (SCC) is based on new types of highly efficient water-reducing admixtures combined with high filler contents, e.g. limestone or special fine-grained sand. The main advantage of SCC is that compacting work with vibrators can be eliminated.

Research into SCC started in Japan in the 1980s. The intention was to manage durability problems caused by insufficient compacting of concrete (Okamura & Ouchi 1999). The first prototype mixes became available in 1988 and made concrete casting possible without vibration. This material was, in fact, named high-performance concrete. However, in Japan, the definition HPC includes self-compacting concrete. Over the past few years SCC has been introduced progressively around the world but the amount of work is still only a fraction of total concrete production. In 1997, SCC amounted to a mere 0.1% of the total production of ready-mix concrete in Japan (Ouchi 1999).

Mizobuchi *et al.* (1999) have described SCC as one of the most innovative developments in the field of concrete technology. Byfors (1999) discusses the use of SCC in the context of the industrialisation of cast in situ concrete, which eliminates compaction work. There are many advantages in using SCC, not least the improved work environment. The elimination of vibration work leads directly to a reduction in manpower on job sites. It accelerates the production process and improves quality, durability and reliability of concrete structures, all of which generate cost savings (Grauers 1999). Smooth, high-quality surfaces can be produced directly without expensive finishing work, that is often needed when casting concrete traditionally. Also the proportion of heavy work is reduced and job sites can be significantly quieter without the noise of concrete vibrators: this is an advantage both for safety on site and for the neighbourhood.

There are also opportunities for designers. For instance, densely reinforced structures, that are difficult or even impossible to construct using traditional methods, can be achieved with SCC. One example is the design of the Millennium Tower in Vienna, which is described by Pichler (1999) as impossible to build without SCC.

Structural frames – production process-related aspects

There are many obstacles to the implementation of new structural frame techniques for house building. The sector is criticised for the low level of co-operation between actors, lack of knowledge, poor inclination to innovate, unclear responsibility, inflexible roles, conservative decision-making etc. In Sweden, the government's Building Cost Commission (BKD 2000) has criticised the house-building sector for lack of customer orientation, technical innovation, holistic consideration (i.e. integration) regarding design, production and use and co-operation between the actors. Although HPC and SCC were not mentioned directly in reporting, failure to actively consider them in house building tends to uphold the view that the sector is deeply conservative and disinclined to innovate.

Construction process

Within the housing sector, some actors in the production process are seldom involved in planning projects. For example, subcontractors and materials suppliers in most cases do not co-operate with either the architect or structural engineer. The opportunity for different actors to influence the choice of structural frame, as outlined by Öberg & Peterson (2000), is largely dependent upon the form of contract adopted for the project. In Sweden, the production of multi-dwelling buildings is dominated by the use of a form of contract for self-development projects. This form of contract often embodies total concepts, which would promote a higher level of co-operation and feedback compared with a more traditional or general form of contract. On the other hand the concepts can be criticised for being too focused on the production phase and based on company standards while reducing the influence of some actors. This is especially true of architects, who may have to break with their own traditions if they wish to become more active during the construction process.

Research project

Project description and objectives

The aim of the project is to investigate the potential of new concrete materials technology for the more competitive production of cast in situ concrete structural frames in multi-dwelling buildings. Aspects of design and production will be analysed, particularly those relating to function and cost efficiency. Implementation issues for this technology, with regard to organisation and the production process, will be investigated.

The opportunity for improved competitiveness in the design and production of structural frames when using new concrete technology, especially high-performance concrete (HPC), has been investigated. One part of the research project consists of a design parameter study that aims at comparing ordinary concrete to HPC, with respect to structural frame dimensions, spans and thickness and amount of reinforcement. The uppermost characteristics of HPC are high and fast-developed compressive and tensile strength. Also important is the concrete's E-modulus.

Fast strength development and drying properties will be analysed, with the aim of investigating earlier formwork removal, reduced winter concreting problems and reduced drying time – factors that contribute directly to a significant reduction in production time. The use of more rational reinforcement will also be investigated. The potential advantages of self-compacting concrete, such as eliminating the *vibration moment* leading to a better work environment, improved safety and greater cost efficiency, will be quantified. In addition, the use of SCC in complex constructions, containing large amounts of reinforcement, will be analysed.

Implementation issues affecting HPC and SCC will be described. Obstacles such as forms of contract and their effect on decision-making, the roles of actors and level of co-operation, as well as competence and interest on the part of the actors involved will be considered. A total concept will be adopted so that sub-optimisation is avoided. Identified advantages in terms of design, production and function will be examined alongside economic aspects and obstacles to implementation.

Research methodology

Quantitative and qualitative methods are used in this research project. The project began with a study of concrete structural frames in Swedish multi-dwelling buildings (Öberg & Peterson 2000). In order to appreciate the scope of the problem and potential opportunities, quantitative data concerning technical and economic aspects were collected. Interviews were conducted with 10 experts from the construction sector.

Finite element methods (FEM) are used in a design parameter study to investigate the effects of different kinds of concrete types on the design-related performance of structural frames. The PC-based computer program *FEM Design Plate*, developed by Skanska IT Solutions (Skanska 2000), simulates the effects of changed concrete parameters on the amount of plate reinforcement, thickness, spans etc. for different kinds of structural frames. Calculations within the computer program are performed to predict the extent of concrete cracking. Exact amounts of reinforcement are calculated that form the basis for a more competitive use of steel. The results of this parameter study indicate the relationships between concrete parameters, amounts of reinforcement and construction dimensions.

A production study will be added to the design parameter study. Account will be taken of the time factor. This implies a study of other concrete properties, such as creep, drying and strength development. The PC-based computer program *Hett* (Cementa 1997), which simulates the hardening of concrete, and *TorkaS* (Lund 1998), which simulates the drying process, will be used. Field studies will be conducted on the impact of SCC on the work environment and the effect of HPC on winter concreting.

A study of implementation issues, with regard to organisational aspects, will be conducted using other survey methods of enquiry, supplementing the earlier interviews. In terms of economics, estimates will be produced of savings attributed to rationalisation so that they can be compared to the additional cost of new materials. A holistic account will be adopted in order to avoid sub-optimisation.

Research results and industrial impact

Quantification of results

A summary of interviews conducted with experts from the sector has been published (Öberg & Peterson 2000). Most of the persons interviewed found that the co-operation between the many actors involved and their interest in new technology was low. Some considered that the cast in situ concrete production process is not an industrial process and that the technique has attendant disadvantages. The study also shows the effects of different contract models and the roles of actors in the decision-making process (for structural frames). For example, ready-mix concrete suppliers cannot affect the choice of structural frame to the same degree in general contracts, as they can in design and build contracts.

Öberg & Peterson (2000) have also produced a technical description of present techniques and market shares of structural frames in multi-dwelling buildings. The design-related parameter study – referred to earlier – shows that spans for structural frames can be increased with the use of HPC and thicker concrete plates or larger amounts of reinforcement are also possible. In the *serviceability limit state*, especially E-modulus, tensile strength and plate thickness affect the displacement and the spans. In the *ultimate limit state*, the amount of reinforcement is a significant parameter. The design study shows the relationship between concrete parameters, reinforcement and construction dimensions for different kinds of plates.

Tentative results

When adding economic and production-related data to the design parameter study, the results will probably reveal whether the direct costs for HPC can be defended on economic grounds, or simply illustrate the potential of new cast in situ concrete technology. For example, when using HPC it might be possible to increase spans by 25% and then also increase flexibility to support future adaptation and refurbishment. This might be economically beneficial, if regard is taken of reduced drying times from six months to one month, or to the reduction of winter concreting problems and the associated heating costs. Another example is that self-drying HPC leads to better acoustic properties with the same production time as ordinary cast in situ concrete. In other words, a larger concrete plate thickness is possible with the same or even a shorter drying time.

The project will also result in a discussion of implementation issues, such as how to address the lack of knowledge and interest in new technology, alongside problems relating to co-operation and responsibilities.

Implementation and exploitation

The tentative results of this research may show technical advantages from the use of new cast in situ concrete technology, leading to more rational and cost-efficient structural frame production from design, production and function perspectives. The results will probably also describe problems connected to cast in situ concrete production of today and production process-related obstacles to the implementation of new concrete technology. The enormous quantity of cast in situ concrete used worldwide suggests that future rationalised solutions are probably capable of generating large cost savings.

Conclusions

During the past decade extensive international research on concrete materials technology has resulted in the emergence of new competitive materials technology: for instance, high-performance concrete (HPC) and self-compacting concrete (SCC). Despite the apparent advantages concerning the design, production and function of structural frames, these technologies are *not* implemented to any significant extent in the house-building sector. The obstacles to implementation probably comprise low interest and competence regarding new technology, decision-making processes, and the roles of actors. International research into these new technologies focuses mainly on technical aspects covering civil engineering works or high-rise buildings. The opportunity for using new concrete technology in low/medium-rise house building has, in comparison, received little attention.

The research project is investigating the advantages of HPC and SCC in terms of the design, production and function of cast in situ concrete structural frames and describes implementation issues with regard to organisation and production process obstacles. Concrete technology used today in cast in situ structural frames for low/medium-rise house building is often characterised as typically low technology. If new concrete materials technology is to be implemented on a large scale it will be because of a more holistic understanding of the benefits, advantages and disadvantages. In particular, economic benefits will be more obvious if regard is taken of, for instance, rationalised and cost-efficient production, improved working environment, greater user flexibility and better indoor climate, and not simply the direct cost of materials.

References

BKD, Byggkostnadsdelegationen (The Swedish Building Cost Commission) (2000) *Betänkande från Byggkostnadsdelegationen, SOU 2000:44*, Stockholm (in Swedish).

Boverket (1999) *Boverkets Byggregler* (The Boards Building Regulations), Karlskrona: The Swedish Board of Housing, Building and Planning (in Swedish).

Byfors, J. (1999) SCC is an important step towards industrialisation of the building industry, *Proceedings of the First International RILEM Symposium on Self-compacting Concrete*, Stockholm: RILEM Publications S.A.R.L, pp. 15–21.

Cementa (1997) *Hett 97*, PC-based computer program, Stockholm: SBUF, NCC AB, Cementa AB.

Chen, Z. & Wang, D. (1996) Utilization of high-strength concrete in China, *Proceedings of the Fourth International Symposium on Utilization of High Strength/High Performance Concrete*, Paris: Presses de l'ENPC, pp. 1571–80.

Chew, M.Y. (1993) Utilisation of high strength concrete in Singapore, *Proceedings of the Third International Symposium on Utilization of High-Strength Concrete*, Lillehammer: Norwegian Concrete Association, pp. 678–90.

Claesson, C. (1999) *Praktisk tillämpning av högpresterande betong,* Göteborg: Byggmästareföreningen Väst/FoU-Väst (in Swedish).

Elfgren, L., Fagerlund, G. & Skarendahl, Å. (1994) Swedish R&D Program on High-Performance Concrete, *Proceedings of the International Workshop on High Performance Concrete in Bangkok.* Michigan: American Concrete Institute, pp. 247–62.

Fagerlund, G., Gillberg, B., Jönsson, Å. & Tillman, A-M. (1999) *Betong och Miljö,* Stockholm: AB Svensk Byggtjänst, pp. 14–48 (in Swedish).

Grauers, M. (1999) Self compacting concrete – industrialised site cast concrete, *Proceedings of the First International RILEM Symposium on Self-compacting Concrete,* Stockholm: RILEM Publications S.A.R.L, pp. 651–58.

Hallgren, M. (1993) *Use of high performance concrete in load bearing structures – Does high strength bring cost savings?* Stockholm: Royal Institute of Technology, Department of Structural Engineering.

Helland, S. (1996) Utilization of HPC, *Proceedings of the Fourth International Symposium on Utilization of High Strength/High Performance Concrete,* Paris: Presses de l´ENPC, pp. 67–73.

Helland, S. (1999) Introduction of HSC/HPC in the market, a contractor's view, *Proceedings of the 5th International Symposium on Utilization of High Strength/High Performance Concrete,* Sandefjord: Norwegian Concrete Association, pp. 14–17.

Hoff, G.C. (1993) Utilization of high strength concrete in North America, *Proceedings of the Third International Symposium on Utilization of High-Strength Concrete in Lillehammer.* Oslo: Norwegian Concrete Association, pp. 28–36.

Ikeda, S. (1993) Utilization of high strength concrete in Japan, *Proceedings of the Third International Symposium on Utilization of High-Strength Concrete in Lillehammer.* Oslo: Norwegian Concrete Association, pp. 37–44.

Jinnai, H., Namiki, S., Kuroha, K., Kawabata, I. & Hara, T. (1999) Construction and design of high-rise buildings using 100 MPa high strength concrete, *Proceedings of the 5th International Symposium on Utilization of High Strength/High Performance Concrete in Sandefjord.* Oslo: Norwegian Concrete Association, pp. 809–18.

Ljunggren, S. (1995) Tysta huset klart för inflyttning, *Fastighetsnytt nr 16 1995,* Stockholm: Fastighetsnytt Förlags AB (in Swedish).

Lund (1998) *TorkaS 1.0,* PC-based computer program, Lund: University of Lund, Institution of Building Technology.

Mängda (1999) *Marknadsanalys husbyggnad,* Stockholm: Mängda AB.

Mizobuchi, T., Yanai, S., Takada, K., Sakata, N. & Nobuta Y. (1999) Field applications of self-compacting concrete with advantageous performances, *Proceedings of the First International RILEM Symposium on Self-compacting Concrete,* Stockholm: RILEM Publications S.A.R.L, pp. 605–16.

Öberg, M. & Peterson, M. (2000) *Multi-dwelling concrete buildings in Sweden: a review of present technology, design criteria and decision making,* Lund: University of Lund, Department of Building Materials.

Okamura, H. & Ouchi, M. (1999) Self-compacting concrete. Development, present use and future, *Proceedings of the First International RILEM Symposium on Self-compacting Concrete,* Stockholm: RILEM Publications S.A.R.L, pp. 3–14.

Ouchi, M. (1999) Self-compacting concrete development, applications and investigations, *Proceedings of the Nordic Concrete Research Meeting in Reykjavik*, Oslo: Norwegian Concrete Association, pp. 29–34.

Persson, B. (1996) Betong anpassad för ett effektivt byggande, *Bygg & Teknik 7/96*. Stockholm: Förlags AB Bygg & Teknik (in Swedish).

Persson, B. (1998) Self-desiccation and its importance in concrete technology, *Nordic Concrete Research*, **21**, Oslo: Norwegian Concrete Association.

Pichler, R. (1999) The use of SCC for building the Millennium Tower in Vienna, *Proceedings of the First International RILEM Symposium on Self-compacting Concrete*. Stockholm: RILEM Publications S.A.R.L, pp.729–32.

Russel, H.G. & Fiorato, A.E. (1994) High-strength concrete research for buildings and bridges, *Proceedings of the International Workshop on High Performance Concrete in Bangkok*, Michigan: American Concrete Institute, pp. 375–92.

Skanska (2000) *FEM- Design Plate 3.50*, PC-based computer program, Malmö: Skanska IT Solutions.

Swedish Building Centre (2000a) *Handbook of High Performance Concrete*, Stockholm: AB Svensk Byggtjänst (in Swedish).

Swedish Building Centre (2000b) *High Performance Concrete Structures – Design Examples*, Stockholm: AB Svensk Byggtjänst (in Swedish).

Walraven, J. (1993) High strength concrete: a material for the future? *Proceedings of the Third International Symposium on Utilization of High-Strength Concrete in Lillehammer.* Oslo: Norwegian Concrete Association, pp. 17–27.

Walraven, J. (1999) The future of high strength/high performance concrete, *Proceedings of the 5th International Symposium on Utilization of High Strength/High Performance Concrete in Sandefjord*. Oslo: Norwegian Concrete Association, pp. 25–36.

Walraven, J. (2000) How can we shorten the bridge between design and material engineers, *Proceedings of the Nordic Concrete Research Meeting in Reykjavik*. Oslo: Norwegian Concrete Association, pp. 1–3.

Chapter 11

Competitiveness in the Context of Procurement

Fredrik Malmberg

Introduction

The last decade has witnessed significant changes to the construction process and, in particular, the ways in which buildings are procured. Structural changes have occurred in the sector as governments and clients modified both their behaviour and their contractual relationships. Subsidies for housing construction have been largely phased out in many countries and demands for more sustainable social structures have increased. Tougher demands have been placed on all actors in the construction process. These demands can be summarised as follows.

- Costs, both long and short term, together with investment and lifetime costs, must be reduced.
- Resources must be utilised more efficiently.
- Opportunities for the client and the end user to influence quality directly (and the utilisation of resources indirectly) must be improved.
- Changes in the construction process to support the goals of sustainability must be improved.
- Buildings must be sound and durable.

New forms of co-operation among the actors in the construction process have been developed to satisfy the above demands. However, the new procurement systems that are based upon these forms of co-operation present problems for purchasing within the public sector. For example, the custom of accepting the lowest tender and ignoring the rest has to be replaced with more rational behaviour, predicated on the understanding that price alone cannot be an indicator of quality or performance. On the contrary, tenders received below what might be deemed to be the most economical price invite poor performance, as contractors seek to *claw back* their costs. In such circumstances the idea that there might be co-operation is swiftly replaced by confrontation and adversarial relationships.

The nature of today's competitive marketplaces has to be understood if tenders are to be prepared and subsequently evaluated properly. Awareness of prevailing and shifting market conditions are fundamental components of the dynamic environment within which tenders are prepared. In addition, current models of tendering and tender evaluation must adapt or be replaced by mechanisms that are compatible with newer forms of procurement, whether they are based on co-operation or, even, some other ideal.

Against this background, the research described in this chapter considers the opportunities for improving the competitive situation for which new forms of tendering and co-operation are being implemented. The overall aim of the research is to develop competition in new forms of responsibility for tendering in new housing and commercial redevelopment. The emphasis of the research is upon evaluating what are today's competition characteristics, the direction of competition and the means for measuring competition.

State-of-the-art review

The research has adopted a broad view of competition, as opposed to considering discrete aspects of it. This holistic approach allows other considerations to be brought into view, that can directly influence or be conditioned by competition. Examples of the former include the labour market and the latter include product and process development. In particular, the research project focuses on questions of competition where new forms of co-operation have been introduced for new housing and commercial redevelopment projects. Other types of construction may, however, be studied.

Competition in tendering for construction contracts

The client (or other person or body responsible for initiating a building project) faces a number of options for organising the project and engaging contractors. One option is, of course, to prepare an overall plan and appoint a designer and then invite tenders from a select list of construction companies once the design is practically complete. The traditional approach is sufficiently understood to avoid elaboration here. Another option is where several (or many) contracts are let, with the contractors responsible for their own defined area, under the overall management of the client, similar to the construction management approach in the UK. In Sweden, this is known as a divided contract (Söderberg & Hansson 1999). There is also an intermediate option that is called the co-ordinated standard approach, which is close to the UK practice of management

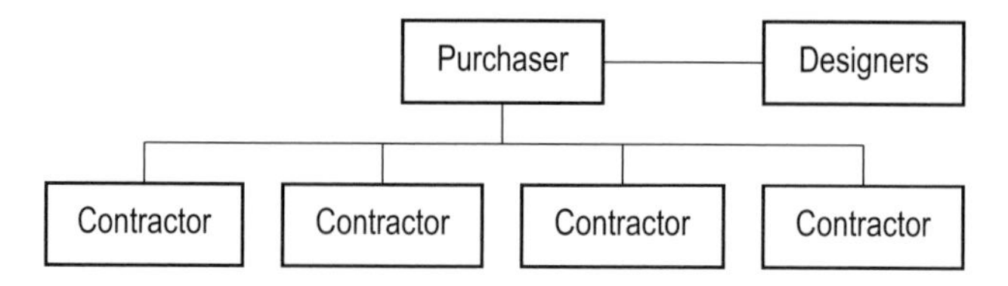

Fig. 11.1 Divided contract approach (after Nordstrand 1990).

contracting. A further option is that of design and construct, that in Sweden is favoured by clients who have limited competence to organise procurement and the tendering competitions that are part of it. Each of the approaches is now described in a Swedish context.

There are many similarities with their international equivalents. After a short explanation of each, their strengths and weaknesses are compared. These will form a basis both for developing new procurement systems and as a guide to the selection of the systems used today.

Divided contract approach (Fig. 11.1)

In the divided contract approach, different contractors are appointed by the client for the individual parts of the project. For each part, the client has to make enquiries, select the most suitable for inclusion in a shortlist, evaluate the tenders and sign a contract with the one that is the most favourable. These contractors are not subject to agreement between themselves, but their work has to be co-ordinated with respect to the project's schedule and logistical considerations. According to *AB92*, common contract conditions published in 1992, the client has the responsibility for co-ordination, but this can be transferred to a main contractor through an agreement. The conditions relating to contract administration clarify those aspects for which another is responsible and the cost involved.

Examples of contractors in this and subsequent figures would include earthworks, structural frame, mechanical work, electrical work and fitting out. The divided contract approach places significant demands on the client to co-ordinate the project.

Standard approach (Fig. 11.2)

As in the divided contract approach, the client takes responsibility for planning the project until the point at which sufficient documents are ready to enable construction to go ahead. The difference in the standard approach is that the results of the planning processes are consolidated in the tender documents to

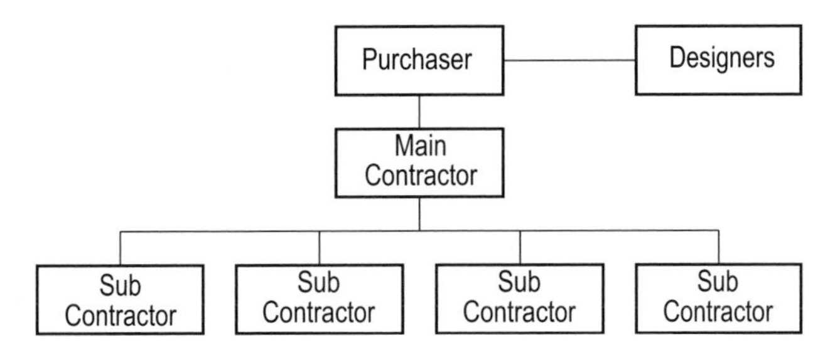

Fig. 11.2 Standard approach (after Nordstrand 1990).

include the whole of the construction works. The client has just one agreement with a main contractor and, therefore, just one party with whom to collaborate during the construction period. The main contractor has responsibilities for co-ordination and for engaging subcontractors.

Co-ordinated standard approach

In this approach the client acts in a similar way as in the divided contract approach and engages the different contractors. Later, one contractor is designated as the main contractor, takes responsibility for the remaining contracts and co-ordinates the project. The advantages of this form of approach are that the client gets a better idea of the different costs within the project, and at the same time does not have the responsibility for co-ordination on site.

Design and construct approach (Fig. 11.3)

The design and construct approach in Sweden is sometimes also referred to as a turnkey project or package deal. In this approach, the contractor is responsible for both design and construction. The client specifies the functional, quality and other demands for the project in detailed tender documents.

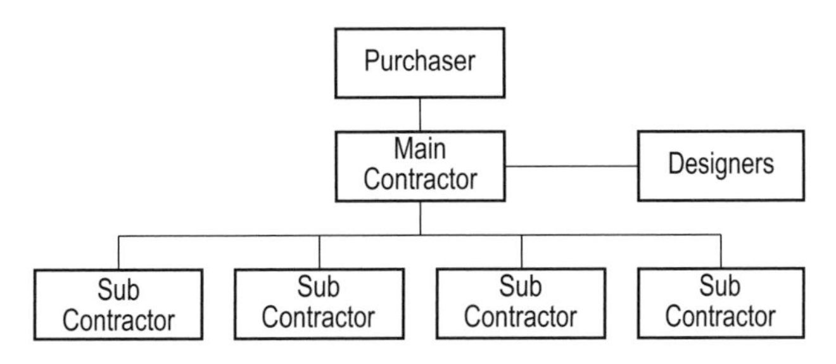

Fig. 11.3 Design and construct approach (after Nordstrand 1990).

An advantage of this approach for the client is that several solutions can be obtained from which to choose the most suitable. Moreover, contractors are provided with an incentive to be innovative. Another positive aspect is that the planning of the project can take less time. Disadvantages of this approach are that it can be difficult to keep track of production costs, and future operating costs may be hard to predict. Another disadvantage is that inviting detailed proposals from several contractors, where only one can win, is clearly wasteful and can indirectly lead to higher prices for the client.

Performance contracting

Performance contracting is a further development of the design and construct approach with the addition of a responsibility for the serviceability of the building. This approach was introduced in Sweden in the beginning of the 1980s on the back of large investments in infrastructure. It strives for early construction starts and is

> 'an approach where quality is controlled by functional demands in the final product that are measurable and where the contractor is responsible for the correct functioning and serviceability [of the building] over several years' (Olsson 1993).

The performance contracting approach is considered to provide an incentive for greater innovation in terms of technology development. Other positive effects are that it is profitable for a contractor to think about the quality and take a more holistic view since there is a contractual responsibility for operations and maintenance costs during the occupancy phase. In order to achieve the prescribed performance, it is necessary to have detailed tender documents that should include the following parts:

- specified demands, and methods for the measurement of their achievement;
- quality thresholds;
- cost limits;
- procurement arrangements;
- feedback during the construction period; and
- handover and commissioning.

The approach assumes that it is possible to identify and agree on key performance indicators that are easy to measure.

PPP – public-private partnerships

When governments are unable or unwilling to fund large building and major infrastructure projects, the private sector can be involved to share the risks and rewards. Public-private partnerships, of which the UK variant is the Private Finance Initiative (PFI), are increasingly seen to offer benefits beyond simply substituting one source of finance for another. However, their popularity is not universal as we see from Chapter 13, *Public-private partnerships – conditions for innovation and project success.*

There are many manifestations of a PPP project; for example, build, operate, transfer (BOT); build, own, operate, transfer (BOOT); and design, build, finance, operate (DBFO); but only three possible kinds of arrangement within the definition of PPP.

- *Financially free-standing projects* – all the costs for the project will be charged to end users, such as a toll road/bridge etc.
- *Services sold to the private sector* – costs for the project are covered through charges for the services to be paid upon their delivery, as in the case of schools, student accommodation, hospitals, prisons etc.
- *Joint ventures* – project costs are recovered partly from end-users and partly through subsidies from public funds.

A PPP project has a long duration and includes all phases of the product's life cycle. When the contractor (acting alone or in consort with others) has to operate the project and generate the revenue from it, there is a tendency to adopt longer term, more holistic thinking. Good projects are those that perform both technically and financially. The aim is not so much about creating an asset as about providing a service. Much of the risk within a project is transferred from the government to private sector actors, on the understanding that the latter are better able to manage risk. However, not all projects will be successful as PPPs. Unrealistic allocation of risks has been a reason for the failure or slow development of PPP projects in some countries.

Summary

Table 11.1 presents a brief summary of the different procurement systems covered thus far, with the exception of PPP, from the client's perspective. The summary incorporates seven different criteria that are used to judge the suitability or otherwise of the respective procurement systems. The primary objective has been to examine characteristics of the systems in a comparative rather than in an absolute sense. A secondary objective is to form some basis for comparison that would enable a client to identify the most appropriate option.

Table 11.1 Comparison of procurement systems.

Approach/contract Aspects	Divided contract approach *Delad entreprenad*	Standard approach *General entreprenad*	Co-ordinated standard approach *Samordnad general entreprenad*	Design and construct *Total entreprenad*	Performance contracting *Funktions-entreprenad*
Focus on price	5	3	4	3	3
Focus on quality	3	3	3	4	4
Focus on time	2	4	3	3	3
Use of soft parameters	3	2	3	2	5
Support of competitiveness	4	3	4	3	4
Overall benefits	2	3	3	5	5
Support for innovation	2	2	2	4	5
Totals	21	20	22	24	29

Note. Scores of 1 are recorded where there is felt to be *no communication with other actors on the same project; no solutions in common between the actors; no life cycle thinking; and no consideration of the interaction between different materials/equipment etc.* Scores of 5 are recorded where there is felt to be *good communication with other actors on the same project; solutions in common between the actors, i.e. solutions made through dialogue before the project on site starts; life cycle thinking; and good interaction between the different materials/equipment etc.*

Each criterion is scored between 1 and 5 inclusive. Scores are bound to be subjective, but should not be seen as static: note that weightings have not been introduced, although could be to emphasise a particularly important criterion or requirement – see Table 11.2 below and later section. The procurement system attracting the highest sum for the selected criteria should be investigated further for its suitability.

Evaluation of tenders

Changes in demand on the building sector are expected to take place without a corresponding decrease in the number of firms competing, and without a reduction in competitiveness. Competition in itself is seen as a driver for change. For procurement in the public sector, legislation governs the approach to tendering and tender evaluation. In recent years, the evaluation of tenders has shifted from being based solely on price to the inclusion of other factors. It has become increasingly common to use models for evaluating tenders that include soft parameters such as competence, quality and environmental considerations (Hatush & Skitmore 1998). Nevertheless, the evaluation of tenders for both housing projects and redevelopment projects is still entirely dominated by evaluation on the basis of price. Models used in tender evaluation are rarely based upon competence and performance. However, methods have been developed that emphasise other factors and have been used to pre-qualify contractors (Wong *et al.* 2001). A case-based pre-qualifying programme called EQUAL is presented by Ng (2001).

By calling for tenders at an early stage, the contractor acquires many degrees of freedom in a competitive situation, such as stimulating new ideas and potentially innovative approaches. Early tendering often gives contractors the opportunity to use their particular competence and experience, and thus influence quality and costs. It is important also to analyse the variation in tender prices, since a cost over-run could place the client in a very difficult position (Shen *et al.* 2001). In order to make a valid assessment of the budget, it is necessary to choose a suitable model for the price forecasting for the construction work (Flanagan & Norman 1983; Morrison 1984). If industrialised construction is contemplated and new systems have to be mastered by the contractor, it is of the greatest importance that the contractor is involved in the project at an early stage.

In recent years, new forms for evaluating tenders have spread. Different implementations of cost-benefit analysis constitute a first step in this direction (Ng & Skitmore 2001). Cost-benefit analysis is a tool to present the consequences of different solutions. It is about assuming/predicting future costs and income/benefits. Several approaches exist. With the net present value (NPV) method it is easy to compare the sum of the yearly benefits against the investment cost.

The result depends, among other things, upon assumptions for inflation and different interest rates. Estimating the value of the lowest tender also depends upon recent experiences. According to Skitmore *et al.* (1994) the last five years of experience from similar projects has an effect on the estimate.

Sensitivity analysis can be performed on cost-benefit models. Different assumptions are used to make forecasts, from which different scenarios for the project can be produced. The particular matter of contractor competitiveness has been measured using regression analysis (Drew *et al.* 2001). Another approach is to use other kinds of mathematical model to forecast contract prices based upon data from similar projects (Skitmore & Patchel 1990).

Several methods can be used to perform evaluations of tenders. Table 11.2 shows an example for illustrative purposes in which two tenders and four parameters, including the price (Parameter A), have been analysed. The scores for each parameter are recorded according to pre-determined scales and the client assigns any weighting. Higher weighting occurs where a certain factor or factors have to be emphasised. The scores can be scaled in two principal ways. The first method is called weighted scaling, where the total is the score multiplied by its respective weighting. The other method is called multiplied scaling, where the total is the score raised to the power of its weighting.

Weighted scaling

$$\text{Tender 1} = 0.30*4.25 + 0.30*3 + 0.20*5 + 0.20*4 = \mathbf{3.975}$$

$$\text{Tender 2} = 0.30*4 + 0.30*4 + 0.20*3.75 + 0.20*4 = 3.950$$

Multiplied scaling

$$\text{Tender 1} = 4.25^{0.30} + 3^{0.30} + 5^{0.20} + 4^{0.20} = 5.633$$

$$\text{Tender 2} = 4^{0.30} + 4^{0.30} + 3.75^{0.20} + 4^{0.20} = \mathbf{5.654}$$

Table 11.2 The two tenders with their respective scores and weighting.

Parameter	Tender 1	Tender 2	Weight
Parameter A	4.25	4	30%
Parameter B	3	4	30%
Parameter C	5	3.75	20%
Parameter D	4	4	20%

The example above shows that multiplied scaling accentuates the differences between the parameters. The weighted scaling leads to a better result for a tender with a peak in a single parameter (Svensson & Jolsvay 1999).

Competitiveness

Tender competitions have traditionally focused on finding the lowest capital cost instead of the lowest whole life cost. When tenders are judged on the basis of lowest price, there is an obvious danger that creative processes that might lead to the optimal overall solution are not considered. Sub-optimisation is the result. In the future, contractors must be able to compete on the basis of factors other than lowest price (Seeger-Mériaux 1997). It is also important that end users have a bigger say (BKD 2000).

Until the present, the majority of contracts have been procured on the basis of drawings showing general arrangements (i.e. floor plans), technical specifications and Gantt charts. A consequence is that contractors cannot compete on the basis of the technical or aesthetic design of the building, unless they want to lose the competition. Thus, competition has remained at the level of production costs, profit and potential claims. This has meant that profitability stems from being knowledgeable in finding loopholes in the drawings and specifications, rather than from presenting new and better solutions.

The definition of good competitiveness as used in Sweden is based on regulations from 1976 that claimed: 'good competition has been achieved when there are three or more tenders'. In order to purchase land from the city and obtain housing subsidies it was necessary to show that procurement was based on good competition, along the lines of the above definition. These rules and regulations have been superseded, but the sentiment still lives in people's minds. This model does not take into consideration how many actors there are in the market, the spread of tenders, if they are serious, how much share of the market is represented by the tendering contractors or if there are any innovative solutions on offer (Seeger-Mériaux 1997). In order to judge whether there is good competition or not, we need new tools/models that are adjusted to the current situation in both the local and the global market. According to Drew *et al.* (2001), the three important factors considered with respect to contractors' behaviour when tendering are type of client, type and size of the construction work.

Research project

Project description and objectives

In order to be able to satisfy the demands placed on the construction sector in the future, a number of changes must be made to procurement procedures, all of which point to the need for new forms of co-operation among all the actors in the construction process. These new forms of co-operation must encourage innovative processes. The emphasis is on evaluating today's competitive situation, the direction of competition and how to measure it. Thus, it is necessary to ensure that it is competition of the entirety and not parts by themselves.

Research methodology

Different types of procurement systems and their associated evaluation systems for tenders are being studied and analysed with competitiveness under the spotlight. The primary approach is to study practices in different countries and to understand how the lessons learned might have application in Sweden. Of particular interest is how the problem of competition is solved, for example, in connection with partnering in the UK and other countries such as the US, Canada and Australia.

Case studies will form the basis of data gathering and analysis. Questions to be pursued in these studies include those relating to models of tender evaluation and their applicability in the context of new forms of co-operation. The studies will also be directed to evidence of a TQM approach on the part of clients and the use of incentives to guarantee better performance from contractors. The impact of different levels of competition will also be investigated, especially the size of contracts and work packages.

The aim is to find a link between the different procurement systems and the types of competition that can be achieved through their use. An implicit question is *what is competitiveness?*

Impact on the sector

The expectation is that the research will lead to a better understanding of how different forms of procurement might affect the level of competition within the construction sector. Through a system of weighted scaling, such as that shown indicatively in Table 11.2, more informed decisions can be made and the most appropriate procurement system selected for a project. The research also expects to report on the opportunity to improve the competitive climate for industrial building projects, for which new forms of tendering and co-operation have been developed. This is intended to lead directly to the development of a tool by which clients can measure competition. In this connection, a review of different types of tender evaluation method will be undertaken, leading to a better understanding of their effects upon competitiveness.

Conclusions

Competition based on competence as opposed to the lowest tender price is a familiar concern. Many clients, public as well as private, now apply qualitative measures to the evaluation of tenders. As less emphasis is placed on lowest price, other factors have come into focus that contractors, in particular, must respond to properly. Of these, the need to adopt more co-operative relationships is fundamental. Moreover, it could lead to less aggressive competition in the future as clients select contractors whom they know can perform. Arguing over price is gone. Tomorrow, contractors will be expected to show what added value they can bring to a contract. That said, there are still impediments to this new thinking. Behaviour has to change, and do so with a sense of a longer-term interest in the project.

References

BKD, Byggkostnadsdelegationen (The Swedish Building Cost Commission) (2000) *Betänkande från Byggkostnadsdelegationen, SOU 2000:44*, Stockholm (in Swedish).

Drew, D., Skitmore, M. & Lo, H.P. (2001) The effect of client and type and size of construction work on a contractor's bidding strategy, *Building and Environment*, **36**, pp. 393–406.

Flanagan, R. & Norman, G. (1983) The accuracy and monitoring of quantity surveyors' price forecasting for building work, *Construction Management and Economics*, **1**, pp. 157–80.

Hatush, Z. & Skitmore, M. (1998) Contractor selection using multicriteria utility theory: an additive model, *Building and Environment*, **33**, *2/3*, pp. 105–15.

Morrison, N. (1984) The accuracy of quantity surveyors' cost estimating, *Construction Management and Economics*, **2**, pp. 57–75.

Ng, T.S. (2001) EQUAL: a case-based contractor prequalifier, *Automation in Construction*, **10**, pp. 443–57.

Ng, T.S. & Skitmore, R.M. (2001) Contractor selection criteria: A Cost-Benefit Analysis, *IEEE Transactions on Engineering Management*, **48**, pp. 96–106.

Nordstrand, U. (1990) *Byggstyrning (Steering construction)*, Arlöv: Almquist & Wiksell (in Swedish).

Olsson, U. (1993) *Funktionsentreprenad för drift och underhåll av vägar och gator (Performance contracting for operational and maintenance for roads and streets)*, Luleå: Luleå University of Technology (in Swedish).

Seeger-Mériaux, A. (1997) *Konkurrens vid upphandling av vägar, beläggningar och drift. (Competition for procurement for roads, paving and operation),* Lund: Lund Institute of Technology (in Swedish).

Shen, L.Y., Fisher, N. & Sun, C.S. (2001) An Analysis of the Distribution of Cost Variance for Building Projects, *Construction Research,* **2**, pp. 35–40.

Skitmore, R.M. & Patchel, B.R. (1990) *Quantity Surveying Techniques – New Directions,* Brandon P.S. (ed.), Oxford: BSP Professional Books, pp. 75–120.

Skitmore, M., Stradling, S.G. & Tuohy, A.P. (1994) Human Effects in Early Stage Construction Contract Price Forecasting, *IEEE Transactions on Engineering Management,* **41**, *1*, pp. 29–40.

Söderberg, J. & Hansson, B. (1999) *Byggprocessen* (The construction process); Lund: Lund Institute of Technology (in Swedish).

Svensson, H. & Jolsvay, V. (1999) *Upphandling med mjuka parametrar – En analys av ett upphandlingssystem* (Procurement with soft parameters – an analysis of a procurement system), Lund: Lund Institute of Technology (in Swedish).

Wong, C.H., Holt, G.D. & Harris P.T. (2001) Prequalification Criteria: A Survey of UK Construction Practitioners' Opinions, *Construction Research,* **2**, pp. 41–56.

Chapter 12
Encouraging Innovation through New Approaches to Procurement

Kristian Widén

Introduction

The need for forms of co-operation that encourage innovation in order to improve the performance of the product and enhance value for the customer has become very apparent in recent years. It seems that most forms of co-operation do not encourage innovation. Are new forms necessary or can those used traditionally be improved? What can be learned from other industrial sectors, and more importantly why has not more been done to transfer know-how? Other sectors appear to have taken the question of innovation seriously and are active in developing their future products, services and markets to a greater extent than the construction sector.

This chapter begins with an overview of traditional innovation theory. It then introduces a new innovation theory that has been developed to cope with a more complex situation, which many sectors face, than that anticipated by traditional theory. The construction sector will be compared with these theories and key factors will be discussed. The chapter will also describe a research project that is addressing these issues.

State-of-the-art review

Innovation

The reason for any company to innovate is to increase its competitiveness. Either it wants to increase its share of the market, and secure a larger return, or it is simply a question of surviving (Atkin 1999). There are many definitions of innovation and some are presented here:

> 'Innovation is the process through which firms seek to acquire and build upon their distinctive technological competence, understood as the set of

resources a firm possesses and the way in which these are transformed by innovative capabilities' (Dodgson & Bessant 1996 p.38).

'Innovation means the application of new knowledge to industry, and includes new products, new processes, and social and organisational change' (Firth & Mellor 1999 p.199).

'Innovation is when an act, such as an invention or idea, begins to impact on its environment' (Atkin 1999 p.4).

It is important to note that innovations are not representative solely of breakthroughs. Audretsch (1995) states that there is evidence of a majority of commercially significant innovations involving the development, application and re-application of existing knowledge with little or no scientific advantage. Furthermore, they do not necessarily bring about changes to existing routines, which is where problems might lie (Hagedoorn 1989). To summarise, innovation is the process where a good idea or creation of new knowledge concerning a product or process begins to affect its context. The innovation can be anything from a small change in a product to a complete shift in the way a company works.

Much of today's work on innovation theory stems from the work of Schumpeter (1976). He argued that companies needed to be large and have a dominant position in order to innovate. Whether large or small firms are better at innovation has been a question that has been argued over many years (Hagedoorn 1989). Acs & Audretsch (1991) found that whether it is large or small firms that are more innovative depends on how innovation has been measured. There is evidence that small firms do make a large contribution in some sectors (Freeman 1982). There are many other factors, however, that influence the likelihood of innovations, and that more or less stem from Schumpeter's work, or the development of it. Dodgson & Bessant (1996) have identified sixteen factors, of which some are more prominent than others:

- competence and knowledge (Barlow 2000; Dodgson & Bessant 1996);
- communication (Barlow 1999; Docherty & Hardy 1996; Hardcastle *et al.* 1999);
- learning (Atkin 1999; Dodgson & Bessant 1996; Lindbeck & Snower 2000; Lundvall 1992; Vickers & Cordey-Heyes 1999);
- relationship and co-operation with other actors (Atkin 1999; Dodgson & Bessant, 1996; Howells 1999);
- risk capital and reward (Atkin 1999; Dodgson & Bessant 1996; Lazonick & West 1998).

Competence is built on the *knowledge* of the people involved in the innovation process. New knowledge is gained through information that is put into context by knowledge gained already (Polanyi 1974; Mitroff 1998). The quality of information is important, as incomplete or wrong information may result in faulty knowledge (Fransman 1996). It is through *communication* that information is transferred. There are two elements to inter- and intra-organisational communications – the channels for communication, and levels of openness and trust (Barlow 1999). The channels used for communication need to be clear and direct so that the message of the information will not be changed. Communication is not just about gaining and sharing information to build knowledge. It is also about administrative information that needs to be clear; for example, how risk and reward is shared and who is the decision-maker. New knowledge is created continuously both internally and externally, with market conditions changing over time. This demands continuous learning (Dodgson & Bessant 1996). Atkin (1999 p.13) states:

> 'organisations that cannot learn cannot innovate effectively'.

Learning is the process when new knowledge is acquired and competence is increased. It is easy to see the strong connection between competence, knowledge, communication and learning. Learning can and will be difficult, if not impossible, if any of the other factors is lacking.

Today, in most industrial sectors, innovation co-operation is becoming more common. Even large multinational corporations cannot expect to be wholly dependent upon their in-house research and technical resources to maintain their innovative performance (Howells 1999). External collaboration in research is used to complement in-house research activity (Arora & Gambardella 1990; Teece 1986). In this way, it is possible to benefit from other companies' research and to reduce the risk and uncertainty involved (Teece 1986). All kinds of inter-firm alliances, not just for innovation, can be problematic. Failures in inter-firm alliances can be the result of contractual problems (Hakansson 1993), loss of project control, loss of technology to the partner, or lack of marketing skills (Dickson *et al.* 1991). Benefits from this kind of alliance do not come automatically. It requires the commitment of the different actors. As innovation depends on the ability to learn, the different actors must commit to transferring knowledge between themselves (Mintzberg *et al.* 1996; Bronder & Pritzel 1992).

Innovation requires financial commitment to succeed (Lazonick & West 1998). The possible reward in the end must be justifiable for the company to put up the risk capital needed for its financial commitment. The balance between risk and reward must always be tilted in favour of the latter (Atkin 1999).

Traditional theories of innovation mostly assume that work in support of innovation takes place in the same company. As this is not necessarily true, there is a need for innovation theories that deal with today's innovation processes rather than those that deal with yesterday's (Nightingale 1998). This is because work in support of innovation is more commonly the result of different alliances and the growing complexity of the innovations themselves. One theory that contrasts starkly with traditional innovation theories and models is CoPS (complex systems). These are defined as:

> 'High cost, technology-intensive, customised, capital goods, systems, networks, control units, software packages, constructs and services' (Hobday *et al.* 2000 p.793).

There are some characteristics that separate CoPS projects from those found in traditional industrial sectors. They are temporary coalitions of organisations that involve many different suppliers. Usually, end users are far more integrated into the process as the product is designed to fit their requirements. CoPS users often need to learn system design skills in order to define their requirements. The multitude of choices makes it difficult to narrow down design options. There needs to be an awareness of how each small system affects every other system (Hobday 1998). In CoPS projects it is necessary to have innovative non-functional organisational structures to co-ordinate production. This is because uncertain and changing user requirements, as well as development and production, necessitate feedback loops from later to earlier stages (Hobday *et al.* 2000).

In terms of innovation in CoPS projects, the actors involved need to agree on the path of innovation. The development of a CoPS project requires a deep understanding of the limits of and possibilities for the system, as well as the capabilities of partner suppliers. Innovation in CoPS projects takes place within production networks with formal alliances. Strong technical performance by an individual supplier will be rewarded with more orders and a greater share of the market. Users have an important role as it is their requirements and feedback, as to what works and what could be better, that triggers the innovation. Users need to share the risks and rewards with suppliers (Hobday 1998).

Hobday also states that one problem in CoPS projects is the question of learning. Transferring knowledge from one project to another is especially difficult as the team dissolves upon completion.

Innovation in the construction process

Construction innovation in the nineteenth century often resulted from new technology and new ways of organising production, that were adapted from other industrial sectors to replace traditional craft-based methods. The change, though, was not as violent as in other industrial sectors, with the traditional and the new co-existing (Gann 2000). During the early part of the twentieth century, the idea of mass production was implemented in the construction sector mostly among suppliers; for example, the shift from hand-made bricks to machine-made. The ability to innovate was one of the reasons for the dominance of a few large companies and this also reduced the level of competition (Gann 2000).

In the 1950s and 1960s it became more common to use prefabricated components, creating a demand for innovation in organisational methods, tools and techniques. At the same time, the various specialists and contractors became more and more isolated from one another, smaller in operational terms and greater in number, thereby hindering innovation attempts (Gann 2000). This also led to an increasingly fragmented marketplace that continues to the present. Winch (1998) argues that volume production and other implications of Schumpeter's theories have failed. The simple reason is that the volume-manufacturing model is not appropriate for the construction sector.

Construction innovation on the part of the primary actors – designers and contractors –other than in a project environment, appears to be rare (Tatum 1987). Neither the designers nor the contractors have large in-house innovation facilities. Ideas that occur on a daily basis on the construction site and in research and development environments normally, literally, have no place to go (Dulaimi 1995). Slaughter (1993) implies that contractors are able make use of their significant experience and expertise to develop useful, effective and low-cost innovations. While the solutions may not always be the best, they are adequate given cost and time pressures.

Many researchers agree that the level of innovation is too low in the construction sector and that the problem rests within the different forms of co-operation (Atkin 1999; Barlow 2000; Latham 1994; Mohamed & Tucker 1996). Reasons for the low level of innovation in construction are diverse. Separation of design and production can be extremely damaging unless these functions can communicate effectively (Hardcastle *et al.* 1999). The fragmentation and discontinuous, project-based nature of the sector leads directly to problems in communication (Atkin 1999; Barlow 2000). As both the level of learning and the quality of information transfer are dependent on how well the actors communicate, these will also suffer. Anheim & Widén (2001) found that the construction process, as it works today, in many cases actively hinders learning. There are some features, more or less common across the different forms of co-operation, that cause problems (Josephson 1994):

- the large numbers of actors involved in projects;
- changing project constellations;
- strict division of the process into phases; and
- the industry's contracts and business structure.

Traditional procurement with fixed price and specification tends to make actors opportunistic (Buckley & Enderwick 1989), limiting the ability to innovate (Mohamed & Tucker 1996), as well as making little use of the expertise of suppliers (Atkin 1999).

In the literature there are ideas of how problems can be solved. There is an underlying assumption that greater collaboration between the different actors would help overcome the problems in the sector (Latham 1994). Even more importantly, competition based on advanced technology and the ability to innovate requires a long-term perspective. Managers of construction firms, long accustomed to intense demands for short-term performance, have to wait for payback over the longer term if they aim for technological advantage (Tatum 1987).

In order to be able to meet changing market conditions, there is the need for a new approach to relationships between the actors in the sector (Edum-Fotwe *et al.* 1999). Specialist contractors and component suppliers have to be more involved in the construction process than is the case today (Atkin 1999). Making information available on how materials, products and systems perform would support the innovation efforts of designers (Atkin 1999). Lenard & Eckersley (1997) suggest that the level of innovation and the ultimate success of a project are highly dependent upon the provision of detailed and accurate tender documentation and appropriate contractual risk allocation. Once this platform is established the level of innovation that occurs is dependent upon four key factors:

- the client's recognition of the need for innovation;
- contractual incentives to encourage innovation;
- creation of symbiotic learning environment; and
- open communication at all levels.

Lagerqvist (1996) has argued that co-operation based on design and build contracts relies on performance requirements that directly stimulate interest in product and process development. Although there is no hard evidence to this effect, there is evidence of enhanced knowledge transfer between the different actors (Hansson 1995). As learning and knowledge transfer between actors is one important step towards an innovative process, this is a step in the right direction. Partnering has also been said to enhance learning and knowledge transfer (Barlow *et al.* 1998). In addition, partnering has been shown to result in technical and process innovation (Barlow 2000). Partnering is, however, neither a procurement form nor a contractual agreement and does not create any legally enforceable rights or duties (Edelman *et al.* 1991).

Future possibilities for the construction sector

The above proposals have more in common with CoPS theories than with traditional innovation theories. Both forms of co-operation – design and build, and partnering – have characteristics that are similar to theories inherent in CoPS. As complex systems and often one-off products too, they have much in common with construction activity. From the theories underpinning CoPS, some issues that need to be addressed in the construction sector can be distinguished:

- deeper understanding of the systems of construction;
- deeper understanding of the capabilities of the actors;
- more involvement of suppliers;
- skilled and involved clients;
- structured and clear feedback loops; and
- clear sharing of risk and reward by the different actors.

Most of these factors are the same as, or aim to achieve, the factors recognised in the literature as prerequisites for innovation. Learning and communication are the factors that are most important to address. In traditional innovation theory they are regarded as important and in the theories of CoPS they have been recognised as problem areas.

Research project

Project description and objectives

The research project aims to develop new forms of co-operation in the construction sector in order to provide greater encouragement for innovation during the construction process than is possible under traditional forms of co-operation. This co-operation between the client, designers and contractors and between contractors and subcontractors is brought into focus at an early stage, along with the different nature of responsibility that each relationship brings.

The results of the research project are intended for use as a base for the continued development and application of new forms of co-operation. They are also intended to indicate the fresh opportunities that the new forms provide and, furthermore, they are to be used as a base for future research projects.

Research methodology

A system theory approach will be adopted as the main method, as co-operation between different actors and the innovation process itself can be seen as systems. In a construction project, many different actors and stakeholders influence the outcome of the project to greater or lesser extents. The innovation process, especially when more than one company is involved, includes many different actors, and their actions will directly affect its success. If one component fails it will negatively affect the other components and, perhaps, the end result. When faced with this situation, a system theory approach is helpful, as a system is a number of components and the relationship between them (Arbnor & Bjerke 1977). A system can be open or closed. A closed system does not take into account the relationship of the surroundings to the system. In this case the construction process is heavily influenced by its environment; it cannot be studied without its context, and so an open system has been chosen.

According to systems theory, every component in a system can be described as a subsystem (Arbnor & Bjerke 1977); for example, the construction process system is built by the different actors or stakeholders as a set of components and each actor system contains different people to take on the role of the components. The research will have limited interest below the level of the different actors as components.

Literature studies and different forms of qualitative studies – for example, case studies with interviews, questionnaires and observations – will be used to provide input to the system and to validate theory with reality.

Research results and industrial impact

Quantification of results

The research includes a deeper study of learning in organisations (Anheim & Widén 2001). In that study, it was found that the application of learning organisation theory would benefit a construction project organisation. Four areas have been identified as crucial to creating a learning organisation: a set of common goals and visions, teamwork, dialogue within the organisation and a complete overview of the organisation. However, the present construction process also needs to change in order to benefit fully from ideas and theories. Moreover, the latter must be adapted to the specific conditions of the construction project organisation for implementation to be successful.

Conclusions

The lack of communication and learning inherent in traditional approaches to the construction process impacts on the extent of innovation that is possible. New methods and technology do not always find their way into the next project that could benefit from them. The low involvement of suppliers and subcontractors is held partly to blame and is a problem that is unlikely to be overcome until changes are introduced to the process. A partial solution to this problem is to create a construction process that encourages and supports innovation. There is the need for further research to identify the means for integrating the innovation process within the construction process. That research has to adopt a standpoint based upon theories of innovation in order to define what the process is expected to deliver. The theories underpinning CoPS could be expected to contribute to a breakthrough in construction innovation for several reasons, including a deeper understanding of the systems of construction and of the capabilities of the actors, greater involvement of suppliers, skilled and involved clients, structured and clear feedback loops and clear sharing of risk and reward by the different actors.

References

Acs, Z. J. & Audretsch, D.B. (1991) Innovation and technological change: an overview. In: Acs, Z.J. & Audretsch, D.B. (eds.), *Innovation and technological change – an international comparison*, Ann Arbor: The University of Michigan Press, pp. 1–23.

Anheim, F. & Widén, K. (2001) Learning organisations in the Swedish construction area, in *Construction Economics and Organization – 2nd Nordic conference*, Gothenburg: Chalmers University of Technology, Department of Building Economics and Management, pp. 259–66.

Arbnor, I. & Bjerke B. (1977) *Företagsekonomisk metodlära*, Lund: Lund Institute of Technology (in Swedish).

Arora, A. & Gambardella, A. (1990) Complementary and External Linkages: The Strategies of the Large Firms in Biotechnology, *Industrial Economics*, **38**, pp. 361–79.

Atkin, B.L. (1999) *Innovation in the Construction Sector*, ECCREDI Study, Brussels, Directorate-General Enterprise, Commission of the European Communities.

Audretsch, D.B. (1995) *Innovation and Industry Evolution*, Cambridge, MA: MIT Press.

Barlow, J. (1999) *Partnering, lean production and the high performance workplace*, Unpublished working paper.

Barlow, J. (2000) Innovation and learning in complex offshore construction projects, *Research Policy*, **29**, pp. 973–89.

Barlow, J., Jashapara, M. & Cohen, M. (1998) Organisational Learning and Inter-Firm 'Partnering' in the UK Construction Industry, *The Learning Organization Journal*, **5**, 2, pp. 86–98.

Bronder, C. & Pritzel, R. (1992) Developing strategic alliances: A successful framework for co-operation, *European Management Journal*, **10**, *4*, pp. 412- 20.

Buckley, P. & Enderwick, P. (1989) Manpower management, in Hillebrandt, P. and Cannon, J. (eds.), *The Management of Construction Firms*, London: Macmillan, pp. 108-27.

Dickson, K., Lawton Smith, H. & Smith, S. (1991) Bridge over Troubled Waters? Problems and Opportunities in Inter-firm Research Collaboration, *Technology Analysis and Strategic Management*, **3**, *2*, pp. 143–56.

Docherty, D. & Hardy, C. (1996) Product Innovation in Large Mature Organisations: Overcoming Innovation to Organisation Problems, *Academy of Management Journal*, **39**, pp. 1120–53.

Dodgson, M. & Bessant, J. (1996) *Effective Innovation Policy: A New Approach*, London: International Thomson Business Press.

Dulaimi, M. (1995) The challenge of innovation in construction, *Building Research & Information*, **23**, 2, pp. 106–9.

Edelman, L., Carr, F. & Lancaster, C. (1991) *Partnering*, IWR Pamphlet-91-ADR-P-4, Ft. Belvoir, VA: Institute of Water Resources, US Army Corps of Engineers.

Edum-Fotwe, F.T., Thorpe, A. & McCaffer, R. (1999) Organisational relationships within the construction supply-chain, *CIB W55 & W65 Joint Triennial Symposium*, Cape Town, pp. 186–94.

Firth, L. & Mellor, D. (1999) The Impact of Regulation on Innovation, *European Journal of Law and Economics*, 1999, pp. 199–205.

Fransman, M. (1996) Information, Knowledge, Vision and Theories of the Firm, in: Dosi, G., Teece, D.J. & Chytry, J. (eds), *Technology, Organization and Competitiveness*, Oxford: Oxford University Press, pp. 147–92.

Freeman, C. (1982) *The Economics of Innovation*, London: Pinter.

Gann, D. (2000) *Building innovation – complex constructs in a changing world*, London: Thomas Telford.

Hagedoorn, J. (1989) *The Dynamic Analysis of Innovation and Diffusion*, London: Pinter.

Hakansson, L. (1993) Managing Co-operative R&D: Partner Selection and Contract Design, *R&D Management*, **23**, pp. 273–85.

Hansson, B. (1995) *A Motorway as a Functional Contract*, Lund: Lund Institute of Technology.

Hardcastle, C., Langford, D.A., Murray, M.D. & Tookey, J.E. (1999) Re-engineering the building procurement decision making process, *CIB W55 & W65 Joint Triennial Symposium*, Cape Town, pp. 265–72.

Hobday, M. (1998) Product complexity, innovation and industrial organisation, *Research Policy*, **26**, pp. 689–710.

Hobday, M., Rush, H. & Tidd, J. (2000) Innovation in complex products and system, Editorial, *Research Policy*, **29**, pp. 793–804.

Howells, J. (1999) Research and Technology Outsourcing, *Technology Analysis and Strategic Management*, **11**, *1*, pp. 17–29.

Josephson, P.-E. (1994) *Orsaker till fel i byggandet*, Gothenburg: Chalmers University of Technology (in Swedish).

Lagerqvist, O. (1996) *Funktionsentreprenad – En modell för upphandling av husbyggnader*, TULEA 1996:26, Luleå: Luleå University of Technology (in Swedish).

Latham, M. (1994) *Constructing the team*, Final Report, London: Stationery Office.

Lazonick, W. & West, J. (1998) Organizational Integration and Competitive Advantage: Explaining Strategy and Performance in American Industry. In: Dosi, G., Teece, D. J. & Chytry, J. (eds) *Technology, Organization and Competitiveness*, Oxford: Oxford University Press, pp. 229–70.

Lenard, D. & Eckersley, Y. (1997) *Driving Innovation: the Role of the Client and the Contractor*, Report No. 11, Sydney: CII Australia.

Lindbeck, A. & Snower, D.J. (2000) Multitask Learning and the Reorganization of Work: From Tayloristic to Holistic Organizations, *Journal of Labour Economics*, **3**, pp. 353–76.

Lundvall, B. (1992) *National Systems of Innovation*, London: Pinter.

Mintzberg, H., Dougherty, D., Jorgensen, J. & Westley, F. (1996) Some surprising things about collaboration – knowing how people make contact makes it work better, *Organisational Dynamics*, **25**, *1*, pp. 60–70.

Mitroff, I. (1998) *Tänk smart*, Oskarshamn: ISL Förlag AB (in Swedish).

Mohamed, S. & Tucker, S. (1996) Options for applying BPR in the Australian construction industry, *International Journal of Project Management*, **14**, *6*, pp. 379–85.

Nightingale, P. (1998) A cognitive model of innovation, *Research Policy*, **27**, pp. 689–709.

Polanyi, M. (1974) *Personal Knowledge – Towards a Post-Critical Philosophy*, Chicago: The University of Chicago Press.

Schumpeter, J.A. (1976) *Capitalism, Socialism & Democracy*, London: Routledge.

Slaughter, E.S. (1993) Builders as Sources of Construction Innovation, *Construction Engineering and Management*, **119**, *3*, pp. 532–49.

Tatum, C.B. (1987) Process of Innovation in Construction Firm, *Construction Engineering and Management*, **113**, *4*, pp. 648–63.

Teece, D. (1986) Profiting from Technological Innovation: Implications for Integration, Collaboration, Licencing and Public Policy, *Research Policy*, **15**, pp. 285–305.

Vickers, I. & Cordey-Hayes, M. (1999) Cleaner Production and Organizational Learning, *Technology Analysis & Strategic Management*, **1**, pp. 75–94.

Winch, G. (1998) Zephyrs of creative destruction: understanding the management of innovation in construction, *Building Research & Information*, **26**, *4*, pp. 268–79.

Chapter 13

Public-Private Partnerships – Conditions for Innovation and Project Success

Roine Leiringer

Introduction

The steady change in the global economy has made national governments more cautious about how they regulate their national debts. Most western countries are finding it increasingly difficult to find a balance between public expenditure, the ability to find funds through fiscal resources and the diverging demands placed on public services. Alternative sources of finance are sought, as well as ways of making public sector services more cost-effective. Partnership solutions between the public and private sectors have emerged as a workable alternative and projects have been realised following the assumption that both sectors have unique skills and characteristics providing them with advantages in undertaking certain tasks. These kinds of projects, known as public-private partnerships (PPPs), are often large and complex, where private sector actors partly or fully undertake the tasks of planning, designing, financing, constructing and/or operating a service usually provided by the public sector. In this way construction companies are entering into agreements that are new to them, bringing with them increased risk exposure and more onerous responsibilities.

Thus, a PPP can be described as:

> an arrangement between public sector and private sector investors and businesses ('the private sector') whereby the private sector on a non or limited recourse financial basis provides a service under a concession for a defined period of time that would otherwise be provided by the public sector. The provision of such service may involve the private sector in the tasks of planning, designing and constructing facilities in order to be in a position to provide the required service.

State-of-the-art review

The term public-private partnership (PPP) has, over the last decade, become politically and socially fashionable. The multitude and diversity of projects that are credited as PPPs are immense and the term is often used to describe a vast range of modern political and financial functions as well as the working arrangements within projects and organisations in multiple areas and industrial sectors. Success is often claimed and several reports show total cost savings of 10–20% over project lifetimes (SO 2000; CIC 2000; Statskontoret 1998). Moreover, the prevailing view is that these kinds of projects provide real incentives and create a business environment that encourages innovation and improved practices in the construction phase (Holti *et al.* 2000; DS 2000; Atkin 1999). It has, for example, been stated by the UK government that:

> 'the search for new opportunities to develop profitable business provides the private sector with an incentive to innovate and try out new ideas – this in turn can lead to better value services, delivered more flexibly and to a higher standard' (SO 2000).

It could be argued that PPPs, regardless of form and size, have added impetus to a change in how construction actors go about their business. Public-private partnerships often consist of an array of systems and subsystems that are subject to a high degree of external influences. They differ from traditional construction projects in the sense that the actors have to take a longer-term perspective. Though it could be claimed that all construction projects have a long-term impact on the surrounding environment, this is even more the case for PPPs, as the success of the constructor in a PPP can only be assessed by the success of the service. In this sense an extra dimension is added to the problem, that of the overall mutual goal of the project. Ultimately, greater consideration has to be taken to align the design, construction and operation phases.

Public-private partnerships

There is no consensus as to the origins of PPP projects. Perhaps this is not so surprising taking into consideration the wide array of projects possible under commonly used definitions. Some claim that the concessions that were common in large parts of Europe during much of the nineteenth and early twentieth centuries were in effect the first PPP projects. Others argue that the concept is a far more recent phenomenon, pointing to the development of project finance techniques for the early North Sea oil projects and various privatisation policies implemented as the means for improving industrial efficiency during that time. In recent history, one of the first widely recognised and documented projects is the Hong Kong Tunnel: a BOT project (see Appendix: *Terminology*, at end of chapter) that was completed in 1972. Since then, several more similar projects have been initiated and completed in various parts of the world although, until recently, not on a continuous basis (Morris 1994; Walker & Smith 1996; UNIDO 1996).

France can be credited for using private finance to upgrade its infrastructure, but the notion of PPP was generally recognised first when the British government launched the Private Finance Initiative (PFI) in 1992. Since then several different contractual arrangements within PPPs have come to the fore such as BOOT, BTO, DBFO and DCMF.

Construction projects are influenced to varying extents by external factors such as laws and regulations, politics and public opinion. This is even more the case for PPPs, and the topic has attracted the attention of a wide range of academic disciplines – see, for example, Montanheiro & Linehan (2000). So far, comparatively little research has been conducted in the field of construction management and in particular on the actual design and construction phases. Instead, most attention has been given to investigations within the fields of economics and political and social science. The findings are all, in their own right, deserving of further consideration. Unfortunately, this cannot be done in such a limited format as this chapter.

Several publications, mainly emanating from the UK, have been issued in the last five years citing varying cost savings and increases in quality generated by projects following the PPP procurement route. In 1998, the UK National Audit Office reported that the first four design, build, finance and operate roads contracts were likely to generate net quantifiable savings of approximately 13% for the state (NAO 1998). A study ordered by the Treasury Taskforce examined 29 private finance projects to reveal an average net present cost saving of 17% (HM Treasury 2000). In contrast, there are also several reports that show increased costs, lower quality products and an overall decrease in service quality: see, for example, Unison (1999) and CUPE (1998).

Caution has to be applied when comparing the findings in the reports, as it is not always a matter of comparing like with like. First of all, projects are not always set up in the same way and cannot at all times be treated as similar. Take, for example, the construction and operation of a toll bridge as compared with the construction and operation of a new high-security prison. Apart from the obvious difference in the contracted service, these projects also differ in several other areas such as regulations and norms that have to be followed (design and operational freedom) and the means of remuneration (the predictability of future revenues). Second, and perhaps more important, the projects are examined in terms of basic parameters. The number of parameters chosen and how they are weighted in comparison to each other will, in conjunction with assumptions, strongly influence the result of the examination. For example, the slightest change in the chosen discount rate will dramatically alter the outcome of any net present value or total life cost calculation. Furthermore, it is important to keep in mind the form of data the report is based upon. It is not uncommon for reports to be based mainly on information gathered entirely from either public or private sector representatives even though it is obvious that the parties have differing views on the value of, say, private sector management.

Most of the claimed cost savings originate from the valuation of risk transfers. For example, ten of the 17% cost savings cited in the above Treasury report are derived in this fashion (CIC 2000). Research on risk (especially risk distributions) is currently being conducted at several institutions covering both assessment of risk and accounting for the transfer of financial risk. However, little research has been done in identifying the cost savings that are to be found outside the scope of successful risk transfers.

In its 2000 study, *The role of cost saving and innovation in PFI projects*, the Construction Industry Council (CIC) identified the role of innovation within construction-based PFI projects. Using a survey based on questionnaires, data were collected from 67 projects targeting clients, client advisers, project managers within the SPV (Special Purpose Vehicle: legal entity that contracts with the public sector client to provide the service) and SPV suppliers. From 108 responses it was concluded that cost savings could accrue from the use of innovative working procedures and new technology. The results show an overall project saving in the region of 5–10% of which the highest average saving could be found from within the construction phase. The savings in the construction costs were also estimated to be 5–10%. It was concluded that the kind of innovation achieved varied depending upon the characteristics of construction. Innovations within civil engineering were mainly technical whereas those in building projects were much less likely to be technologically-based (CIC 2000). The report provides useful insights into what senior managers consider to be key improvement issues in PPPs: it does not, however, deal with implementation issues.

Guidelines and standard documents – see, for example, *The Partnerships Victoria Guidance* (State of Victoria 2001) – are not uncommon. These are, however, created by public sector appointees and they target public sector clients. Although the private sector clearly benefits from these guidelines, little has so far been done to enhance the interests of commercial enterprises/private sector companies in general and construction companies in particular, with a few exceptions – see, for example, CIC (1998).

Innovation

Innovation as a field of study has existed for several decades; indeed, literature can be found as far back as the early twentieth century (Padmore *et al.* 1998).

Primarily, focus has been on the manufacturing sector of national economics (Slaughter 1998). Several definitions have been provided differing mainly in their level of detail. A technical innovation is defined by the OECD (1996) thus:

> 'a technological product innovation is the implementation or commercialisation of a product with improved performance characteristics such as to deliver objectively new or improved services to the consumer. A technological process innovation is the implementation/adoption of new or significantly improved production or delivery methods. It may involve changes in equipment, human resources, working methods or a combination of these'.

On a more general level, Freeman's (1982) definition of an innovation as the actual use of a nontrivial change in a process, product or system that is novel to the institution developing the change, has been commonly accepted. Of significance in this connection is the emphasis on the change being novel to the institution concerned. While this is also true for inventions the two should not be confused. An innovation could very well be an invention but an invention is not necessarily an innovation unless it has actually been used – an innovation does not have to be novel to the existing arts. The ERT (1998) states that innovation should be seen as something greater than merely new technology, science and research, and above all it should not be seen as a strictly economic issue. It should also be considered as a way of organising work and social structures in more efficient and humane ways, making organisations more competitive and the workplace more satisfying. However, for the sector, innovation is, even if it may not be the sole reason, about profit generation. Take away this aspect and there is little point in investments that would merely consume resources without payback.

Regardless of the definition adopted, it has long been common practice to differentiate between kinds of innovations and their impact on their surroundings. In his seminal work from 1934, Schumpeter credited innovations as a way of increasing economic growth and differentiated between five kinds:

(1) introduction of a new product or a qualitative change in an existing product;
(2) process innovation new to an industry;
(3) the opening of a new market;
(4) development of new sources of supply for raw materials or other inputs; and
(5) changes in industrial organisation (Padmore *et al.* 1998).

While there is little point in stating that construction is specific, as this goes for all industries, it is still important to bear in mind that the sector has its particular characteristics. Models created to describe innovation in manufacturing are not likely to be successfully applied to the construction context without a certain degree of modification. There are certain key differences that distinguish construction from manufacturing that have to be taken into consideration. Slaughter (1998) proposed that innovation in construction could be described by a set of models based on two main principles:

(1) The magnitude of the change from the current state-of-the-art associated with the innovation.
(2) Linkages between the innovation and other components and systems.

Five types of innovation are recognised:

- *Incremental innovation* – small changes based on existing technology. Its origin is often to be found within the organisation implementing it.
- *Modular innovation* – a significant change within a component, but one that has little effect on other components.
- *Architectural innovation* – this constitutes a small change in a component but a major change in the links to other components and systems.
- *System innovation* – integrates multiple independent innovations to perform new functions.
- *Radical innovation* – involves a breakthrough in science or technology that could very well change the character of the sector.

Innovations in construction are most commonly of the incremental or modular kind (Koskela & Vrijhoef 2001), meaning that the sources for improvement in construction are most often to be found within organisations that already exercise control over the components and modules/systems. In a recent study, Lenard (2001) shows significant differences between the manufacturing and construction sector's views on competition. Manufacturers perceive their competitors as achieving greater market share. They therefore invest considerable energy in monitoring competitors and encouraging subcontractors to become part of their specific sphere of interest. In contrast, construction actors perceive their counterparts as predictable and regard their main competition as coming from their own clients, suppliers and subcontractors.

A major deficiency in the study of innovations in construction is the mix between organisational and inter-organisational issues. Often it is assumed that the same driving forces that lie behind innovation on an organisational level are also key within unique project organisations. Innovations are too often looked upon through an organisational perspective, i.e. as if the innovation is a result of the work within the organisation, which is in itself a solid entity. The interaction the organisation has with other actors and influences thereof are disregarded (Gann & Salter 2000). Following this logic, the unit of analysis should not be taken away from its context and studied as a unique phenomenon. Furthermore, innovations should not be looked upon as simply one of occurrences.

Innovation is, as mentioned earlier, a change that is novel to the institution developing it. The innovation could therefore be of an intermediate kind, adding to the problem of assessing the significance of the change and the reasons for its adoption. Essentially an innovation could be either proactive or reactive. In work targeting the use of intermediate technology in manufacturing industry, Lissoni (1999) suggests that technical progress is continuous and that adoption of the latest technology can be postponed not only by waiting, but also by adopting intermediate technology. In this context, issues such as supply and demand cannot be discarded.

Research project

Project description and objectives

The research project featured in this chapter was initiated in 1999. The main goal of the project is to clarify and strengthen the role of the construction sector in PPPs. Particular interest is given to the design and construction phases of the projects.

The aim is to determine whether or not the PPP procurement route enhances the possibilities for the construction sector to be innovative and if the way projects are set up enables novel practices and technology to be successfully implemented. Part of the study is also dedicated to identifying the potential for PPP projects in Sweden from selected studies of overseas markets, highlighting key differences that are likely to affect the successful working of such an arrangement. The aim is to pinpoint practices and techniques that work well in practice so that the whole affair becomes more transparent, enabling construction companies to understand and then adopt a best practice approach.

Although transportation infrastructure is excluded from the study, lessons that might also apply in such a context are being considered.

Research methodology

The nature of the target application – PPPs – is such that several considerations have to be taken in establishing the approach to the research. Thus, the research aims to answer two interrelated questions:

(1) Does the PPP procurement route support the implementation of innovative practices and novel technological solutions in construction?
(2) What are the main inhibitors to innovation within PPPs?

There are several proven research strategies that could be applied to this line of research, all of which have their own advantages and disadvantages. Each is a different way of collecting and analysing evidence but the boundaries are not always clear and sharp, and even though each strategy has its own characteristics there are large areas of overlap among them. The main difference between the approaches is the relationship between the breadth and the depth of the underlying study.

Construction projects are generally unique and of a longer duration than the end products of industrial processes. This is especially true for PPPs where the result can only be measured in the terms of the service provided. This, in conjunction with the exploratory nature of research questions, has led to several single-method research approaches being considered and subsequently rejected. These include:

- *Histories*, considered as the preferred strategy in explanatory research when the investigator has virtually no control over or access to events (Yin 1994). The method was disregarded because documentation on construction projects would not suffice in providing relevant information to answer the research questions. Furthermore, considerable difficulty is attached to establishing the objectivity of recorded data.
- *Ethnography* has its roots in anthropology and has been credited as an effective method for gaining insights that enable questions such as: *what, how and why certain events occur?* The method was discarded due to the limited time and resources allocated to the research project. The relevant phases of a PPP project are usually of too long a duration for sufficient data to be collected.
- An *experiment* can be defined as 'a study in which certain independent variables are manipulated, their effect on one or more dependent variables is determined and the levels of these independent variables are assigned at random to the experimental units in the study' (Hicks 1982). In a scientific context, experiments are devised to investigate any relationship between activities carried out and the resultant outcomes. In studying construction projects, it is often very difficult to isolate individual dependent variables.

An approach considered was that of comparative studies of similar projects, undertaken at the same time by similar types of organisations (Fellows & Liu 1997). This approach would fit both of the research questions but it was deemed to be unworkable due to the time constraints of the project. PPP projects are of too long a duration for a purposeful experiment to be conducted and it is nearly impossible to fulfil the requirements of similar projects at the same stage in time.

A *case study* has been defined as something that

> '...investigates a contemporary phenomenon within its real-life context, especially when the boundaries between phenomenon and context are not clearly evident' (Yin 1994).

Multiple sources of evidence are often used, e.g. documentation, interviews and questionnaires. In the case of PPPs, one case study would not be enough to answer the research questions, meaning that a multiple case approach would be necessary. The method of generalisation would then have to be that of 'analytical generalisation' in which a previously developed theory is used as a template. This would allow the acquired empirical results to be compared and both questions to be answered. However, the amount of work would surpass the resources allocated to the project.

The chosen research approach for the project is *multi-method*. This has been claimed to add to the strength of evidence collected (Brewer & Hunter 1989). In this research they cover:

- a wide literature review examining leading academic and technical journals, technical reports, conference proceedings, case studies, the financial/business press and government guidelines;
- the establishment of a reference group consisting of 14 members, all of whom are considered to be experts within their respective fields, to represent the major stakeholders in PPP projects – the group is used as a means for validating the findings as well as generating new insights and knowledge through seminars and workshops;
- semi-structured interviews with senior representatives of organisations involved in the PPP process;
- fieldwork with the researcher taking the role of observer within the BOT unit of a large construction company; and
- a survey targeting individuals representing firms active in the various phases of PPP projects with the emphasis on the design and construction phases.

A survey constitutes the main part of the data collection. Surveys have the advantage of allowing a greater sample of projects to be taken into the research while limiting the human resources committed by the research team. However, the use of questionnaires alone leaves the data vulnerable to bias and reactive measurement effects, and the method does not allow for analysis to be made of the relationships between independent variables and causal relationships. Therefore, multiple sources of evidence are to be used, i.e. documentation, interviews and questionnaires. The method operates on the basis of statistical sampling, as a full population survey is not possible or indeed necessary. It is expected that these actions will reveal valuable information as to which criteria have to be fulfilled in order for innovative approaches to be implemented within PPP projects. Conclusions will be drawn from the results outlining successful practices, the responsibilities of those involved and potential benefits that can be gained. Likewise, attention will be given to failures and the reasons behind them.

Research results and industrial impact

This research project is not about assessing or passing judgement on the rights or wrongs of including the private sector in domains that previously have been operated solely by the public sector. Neither does it deal with issues such as the socio-economic benefits of projects being realised ahead of time. However, these are the kind of issues that tend to dominate the public debate. This ongoing debate is very much impeded by confusion within the ranks of the public sector and the media, as well as within the private sector. So far no real consensus has been reached as to the meaning that is given to the term PPP and acronyms such as PFI, BOT, BOOT, BTO, DBFO and DCMF are used interchangeably (see Appendix: *Terminology*, below). The parties often appear to be discussing different issues and it seems that the public sector is highly influenced by prejudices and political sympathies.

Quantification of results

A position report, based on a literature review *Public Private Partnerships in Swedish Construction* (Atkin & Leiringer 2000a) has been produced, covering the current position of PPP in Europe in general and Sweden in particular. It concluded that PPP projects are ever more common and that a clear trend can be seen, one where governments are showing an increasing willingness to experiment with alternative procurement routes. PPP projects are being carried out or are about to start all over Europe and, although the projects are not in the majority in terms of their size or expenditure, they do represent a considerable volume of construction work.

The projects naturally differ depending on the end product and the given geographical and legal restrictions of the country in which they are realised, but there are also several other key aspects that separate the projects. In order to be able to compare like with like, the foundation of a typology for PPP projects has been created based on published reports, guides and manuals. This work has resulted in the presentation of a number of key areas that have to be taken into consideration when assessing the projects (Atkin & Leiringer 2000b; Leiringer 2001).

Inhibitors to the successful implementation of innovative procedures have been identified during the course of the research, through the means of literature reviews, structured interviews, observational fieldwork within a large contracting organisation's specialist BOT unit and group seminars. These inhibitors take the form of contradictions in the logic between what needs to be in place to enhance innovative behaviour and the way that most PPP projects are set up and managed.

It is expected that the survey will reveal valuable information as to which criteria have to be fulfilled in order for innovative approaches to be implemented within PPP arrangements. The aim is to show how the innovation compares with what is going on in the sector as a whole, and if there are any direct connections to its realisation (and that it took place in a PPP project). The intention is to show if anything in the project set-up or in the way the project was procured enables or inhibits innovation. This information will subsequently be instrumental in the preparation of guidelines for construction sector actors venturing into PPPs.

Implementation and exploitation

Sweden is at a crossroads with fundamental decisions pending at the highest of political levels. Regardless of whatever decisions are taken in the near future there seems to be little doubt that PPP projects will be realised sooner or later. With the driving forces arguably being somewhat different from that of other countries, it seems likely that the concept will be more focused on providing a high-quality service than on getting the projects off the public sector balance sheet. This means that projects and therefore the actors involved will be assessed on their ability to provide something better (in terms of time, cost and quality) than would be the outcome of a traditionally procured project. Evidence from elsewhere shows that this could be done, but so far it has not been achieved on a continuous basis. There are numerous examples of projects that have not been able to fulfil their stated objectives.

Very few Swedish construction companies have a track record in these kinds of arrangements due mainly to the lack of projects. Knowledge of the kinds of arrangements that are conducive to innovation will increase the chances of project success. Furthermore, what is desirable from clients, financiers and other stakeholders may not lead to an optimal solution from a design and construction perspective. Indeed there are potential project set-ups that are far from ideal for construction companies to venture into, both in the sense of taking shares in the SPV as in undertaking actual construction work. It is believed that construction actors would benefit from guidelines and that these would help accelerate learning and enhance the possibility of a successful outcome.

Conclusions

Internationalisation is likely to lead to PPP projects becoming even more common. Much has been said regarding the potential of the PPP procurement route and several benefits are claimed to accrue from its implementation. However, PPPs should not be looked upon as simply a source of work for the construction sector. There exist obstacles and contradictory evidence of the emergence of improved practices. Not all projects are suited to the PPP procurement route and not all PPP projects are an ideal venture for construction companies. Best practice is yet to be established and it is still to be proven whether or not this procurement route lives up to what is claimed.

Appendix: Terminology

BDO: Buy Develop Operate
BLT: Build Lease Transfer
BOO: Build Own Operate
BOR: Build Operate Renewal of concession
BOT: Build Operate Transfer
BOOT: Build Own Operate Transfer
BRT: Build Rent Transfer
BT: Build and Transfer
BTO: Build Transfer Operate
DBFO: Design Build Finance Operate
DCMF: Design Construct Manage and Finance
LDO: Lease Develop Operate
LROT: Lease Renovate Operate Transfer
MOT: Modernise Own/Operate Transfer
OM: Operate and Maintain
PFI: Private Finance Initiative
PPP: Public-private Partnership
ROO: Rehabilitate Own Operate
ROT: Rehabilitate Own Transfer

References

Atkin, B.L. (1999) *Innovation in the Construction Sector*, ECCREDI Study, Brussels, Directorate-General Enterprise, Commission of the European Communities.

Atkin, B.L. & Leiringer, R. (2000a) *Public Private Partnerships in Swedish Construction*, Stockholm: Svenska Byggbranschens Utvecklingsfond.

Atkin, B.L. & Leiringer, R. (2000b) Defining the Concept of Public Private Partnerships, Unpublished report, Stockholm: Royal Institute of Technology.

Brewer, J. & Hunter, A. (1989) *Multimethod Research: A Synthesis of Styles*, Newbury Park: Sage Publications.

CIC (1998) *Constructors' key guide to PFI*, Construction Industry Council, London: Thomas Telford Ltd.

CIC (2000) *The role of cost saving and innovation in PFI projects*, Construction Industry Council, London: Thomas Telford Ltd.

CUPE (1998) *Behind the pretty packaging: Exposing Public Private Partnerships*, [Online] http://cupe.ca/downloads/PPP0898en.pdf [10 March 2001]

DS (2000) *Alternativ finansiering genom partnerskap – Ett nytt sätt att finansiera investeringar i vägar och järnvägar*, Departementsserien 2000:65, Näringsdepartementet, Fritzes Offentliga Publikationer, Stockholm (in Swedish).

ERT (1998) *Job Creation and Competitiveness through Innovation*, The Brussels: European Round Table of Industrialists.

Fellows, R. & Liu, A. (1997) *Research Methods for Construction*, Oxford: Blackwell Science.

Freeman, C. (1982) *The economics of industrial innovation*, 2nd edition, London: Pinter.

Gann, D. & Salter, A. (2000) Innovation in project-based, service-enhanced firms: the construction of complex products and systems, *Research Policy*, **29**, pp. 955–72.

Hicks, C.R. (1982) *Fundamental Concepts in the Design of Experiments*, 3rd edition, New York: Holt–Saunders International.

HM Treasury (2000) *Value for Money Drivers in the Private Finance Initiative* (A report by Arthur Andersen and Enterprise LSE), London: HM Treasury Private Finance Taskforce.

Holti, R., Nicolini, D. & Smalley, M. (2000) *The handbook of supply chain management – The essentials* (C546), London: Construction Industry Research and Information Association.

Koskela, L. & Vrijhoef, R. (2001) Is the current theory of construction a hindrance to innovation? *Building Research & Information*, **29**, *3*, pp. 197–207.

Leiringer, R. (2001) Understanding Public Private Partnerships, *Proceedings 2nd Nordic Conference on Construction Economics and Organization*, Gothenburg, Sweden, April 24–25, pp. 249–58.

Lenard, D. (2001) Promoting the development of an innovative culture through the strategic adoption of advanced manufacturing technology in construction. *Proceedings CIB World Building Congress*, Wellington, New Zealand, April, **1**, pp. 220–29.

Lissoni, F. (1999) *Technological expectations and the diffusion of 'intermediate' technologies*, Working Paper 8, CRIC, Manchester: UMIST.

Montanheiro, L. & Linehan, M. (eds) (2000) Public and Private Sector Partnerships: The Enabling Mix, *Proceedings of the 6th International Conference on Public and Private Sector Partnerships*, Sheffield: Sheffield Hallam University Press.

Morris, P.W. (1994) *The Management of Projects*, London: Thomas Telford Services.

NAO (1998) *The Private Finance Initiative: the first four design, build, finance and operate road contracts*, London: Stationery Office.

OECD (1996) *Proposed Guidelines for Collecting and Interpreting Technological Innovation Data – Oslo Manual*, Paris: Organization for Economic Co-operation and Development.

Padmore, T., Schuetze, H. & Gibson, H. (1998) Modeling systems of innovation: An enterprise-centered view, *Research Policy*, **26**, pp. 605–24.

Slaughter, E.S. (1998) Models for construction innovation, *Construction Engineering and Management*, **124**, 3, pp. 226–31.

SO (2000) *Public-private Partnerships: The Government's Approach*, London: The Stationery Office.

Statskontoret (1998) *Privatfinansiering Genom Partnerskap*, Stockholm: Statskontoret (in Swedish).

The State of Victoria (2001) *Partnerships Victoria Guidance*, Melbourne: The State of Victoria, Department of Treasury and Finance.

UNIDO (1996) *UNIDO BOT Guidelines*, Vienna: United Nations Industrial Development Organization.

UNISON (1999) *Downsizing for the 21st Century*, London: UNISON.

Walker, C. & Smith, A.J. (1996) *Privatized infrastructure: the Build Operate Transfer approach*. London: Thomas Telford Services.

Yin, R.K. (1994) *Case Study Research: Design and Methods*, 2nd edition, Thousand Oaks: Sage Publications.

Chapter 14
Pros and Cons in Partnering Structures

Anna Rhodin

Introduction

Wider use of partnering can be a driving force for development of the construction process. The phenomenon itself is multifaceted and so are the motives for using it. Principles and practices need to be developed step-by-step and adopted at different levels within organisations. Contextual adjustments and a realistic view of expected outcomes, related to investments in terms of money and time, are crucial for the future acceptance of partnering.

The consensus formed from a study of literature on partnering is that, at the broadest level, it is an arrangement whereby client and supplier seek a mutually effective form of association. It can involve a commitment to work closely together for the duration of a single project or for a number of projects. Partnering includes a set of processes to support collaboration and these can be modelled. Some of them are not new for the construction sector. The strength of partnering arrangements includes the flexibility to organise and the opportunity (and importance) of scaling or tailoring its application according to project-specific characteristics.

This chapter aims initially at reviewing the state-of-the-art for partnering in construction. The review demonstrates the comprehensiveness of the concept leading to implications for practitioners as well as researchers concerning partnering in the construction sector. Additionally, the outline of a research project is presented. The research is addressing the question of how formalised and structured partnering enhances or inhibits collaboration in small- to medium-size multi-partner activities. Some tentative results are discussed.

State-of-the-art review

Issues related to partnering in construction represent the main part of the following section, since this is the primary area of research. Also discussed, albeit briefly, are inter-organisational collaboration, group processes and communication.

Partnering in construction

There is a division between those who see partnering as an informal and organic development and those who regard it as something formal and actively engineered. This separation between formal instrumental views and informal developmental views on partnering is reflected also in attitudes towards the role of contracts in such arrangements (Bresnen & Marshall 2000a).

In the following state-of-the-art review, characteristics and attributes related to partnering are broadly examined from informal as well as formal perspectives. A critique on partnering in construction and research on the topic is summarised separately. Results presented later constitute a small part of a broad study of partnering in a Swedish context. In order to form a theoretical framework for discussion of results in the chapter, the final section deals with the selection process, role of contracts and the external environment in partnering arrangements.

Since the late 1980s partnering in construction has drawn a lot of attention from academics and practitioners. Most documented work has emerged from experiences in the US, UK and Australia. One study of partnering research in construction (Li *et al.* 2000) summarises the past 10 years of documented issues published in 29 articles in the highest rated construction management journals. They found four major themes of empirical research: project partnering; examination of a dual relationship; international partnering and special applications of partnering. Non-empirical research and studies focused on types of partnering, partnering models, partnering processes and partnering structures. Half of the papers reviewed covered empirical work.

A few studies are based on large samples of projects. Larsson (1995) investigated the contractor–owner relationship in 280 construction projects to measure the degree of success related to alternative approaches. The criteria used to measure project success were cost, schedule, technical performance, customer needs, litigation avoidance, participant satisfaction and overall results. Differences between informal partnering (77 projects) and more structured and formalised partnering approaches (59 projects) were significant. Larsson supports the idea of a more structured approach to partnering. The impact of improved interaction on objective performance measures is validated in a study of 209 projects, including 63 based on partnering. *Degree of Interaction* (DOI) is used as an objective, quantifiable method of approximating project integration. Partnering and combined projects had significantly better performance than traditional projects in three of the four indicators, namely cost, schedule, modifications and design deficiencies (Pocock & Liu 1997).

Partnering in construction is becoming a well-established way of contracting in the UK, US and Australia, unlike the situation in Sweden. Practical guides to best practice are common in the English-speaking countries. Considerable academic work and practice on partnering has been conducted in the UK during the 1990s. Guides to so-called *second and third generation* partnering have been produced (Bennett & Jayes 1998). A comparison of experiences on partnering between Sweden and UK indicates that, in the former, partnering moves between the *first and second generation* depending on the aspect analysed (e.g. processes, clients and teams). A couple of studies have been carried out in Sweden with emphasis on cost reduction through common goals, integrated organisations and analysis of work processes (Andersson & Borgbrant 1998; Persson 1999).

Behavioural aspects of relationships

Partnering places special emphasis on front-end decision-making, planning and execution phases of projects, as well as the rapid integration of organisations with different knowledge bases (Barlow 2000). The emergence of trust is critical in influencing the scale and scope of knowledge transfer between organisations (Barlow *et al.* 1997). There are similarities between partnering as a process for effective collaboration and the ideas of learning organisations. DeVilbiss & Leonard (2000) have created an implementation model for a learning organisation where partnering is the foundation.

Many authors identify mutual trust as a key characteristic in successful partnering. Barlow *et al.* (1997) emphasise the characteristics of individuals and their openness and willingness to accept and share responsibility when it comes to mistakes. Furthermore, the presence of more open and flexible communications for the development of trust is strongly emphasised. Reduced transaction costs through social capital and trust have been argued by Williamson (1975). Lazar (2000) uses game theory to explain how trust-based relations develop, and demonstrates that a mixed strategy produces higher scores for all players than either unconditionally collaborative (co-operative) or conditionally competitive (adversarial) behavioural strategies. Lazar distinguishes between trust-based and reciprocity-based relationships in partnering arrangements where the latter is the most fragile.

Perspectives of the Eastern culture and how it can contribute to project success through partnering is given by Liu & Fellows (2001). In Eastern business, personal integrity is the primary consideration, while Western business emphasises legal formalities in the shape of contracts. Management of conflict through self-cultivation and goal setting that yields goals sufficiently attractive to all participants are suggested as essential issues for the adoption of the partnering process. Culture manifests itself in behaviours that are underpinned by values. The importance-hierarchy of values to the self dictates one's behaviour. According to goal setting theory, rewards affect the degree of commitment to goals. If individuals benefit from trusting behaviours they will continue their self-cultivation. In such a holistic approach to goal attainment in partnering, the emphasis on a physical structure to dictate the partnering process will miss the point.

An experimental, large-scale survey on the contractor and designer relationship illustrates that the areas most likely to contribute to disintegration seem to be those that have the greatest link to the organisation. The designers had an apparent lack of concern for general contractor profitability. Architects, in particular, tend toward a traditional arm's length relationship, whereas contractors favour integration. The results imply, if they hold in practice, that improving project performance through integration is not without significant difficulties (Puddicombe 1997). Kadefors (1999) discussed five identified driving forces and dispositions affecting the client, design team members and the contractor in negotiations centred on changes (variations). The fairness constraint, economic interests, status aspects, intuitive information-processing biases and civic spirit together make conditions of interaction less transparent and predictable. There is clear evidence in the studies of complexity when it comes to behaviour. An awareness of these underlying factors is fundamental for progress in partnering research and for partnering implementation.

Culture, as a theme, receives relatively little attention in partnering research although it is often stated as important. Bresnen & Marshall (2000a) observed the simplistic way in which partnering has been examined. They mentioned the existence of sub-cultures and the complex dynamics associated with changes as two examples of cultural difficulties. A cultural fit between collaborating firms can provide a basis on which mutual confidence and trust can develop (Faulkner & De Rond 2000).

Tools and techniques suggested for partnering

The use of different formal tools and techniques is often proposed to support and develop partnering processes. Project managers, however, would benefit from an understanding of the underlying factors that the tools address, as well as the tools themselves (Puddicombe 1997). Results from the study by Larsson (1995) of 280 construction projects emphasise the value of investing in more structured partnering approaches. Superior results in terms of controlling costs, achieving technical performance and meeting customer needs and in the overall results were reached in the formalised partnering projects. No significant differences were found in the study between formal and informal partnering projects related to meeting schedules and avoiding litigation.

The formation of a partnering charter has been seen as the single most beneficial tool in developing a co-operative partnering relationship (Thompson & Sanders 1998). Common goals for involved parties are agreed in a document at an early stage of a project. In order to achieve alignment of the team, it is important that goals are really shared and clearly understood by everybody. An important instrument in the process of building trust and aligning parties with different perspectives is teambuilding (Barlow *et al.* 1997). Formal teambuilding with external facilitators did not seem to be evident in the partnering cases studied, neither was teambuilding repeated at key project stages. Formation of charters and teams are activities that are expected to be managed in the form of meetings or workshops for one or two days at which the key players are in attendance. Follow-up workshops for evaluating progress and induction workshops, when someone is appointed later in a project, are recommended in practitioners' literature, alongside final workshops for reviewing project performance (Bennett & Jayes 1995).

In observations of case studies in partnerships and alliances, limitations in the use of incentives are verified. The implication is that systems for enhancing motivation need to be very carefully designed. Motivation and commitment operate at different levels of analysis with respect to the organisation and the individual. Participant evaluation of rewards, expectation of performance and perception of equity are highly subjective and that they may differ must be understood. Intrinsic as well as extrinsic rewards influence motivation and commitment (Bresnen & Marshall 2000b).

Pietroforte (1997) noticed a dislocation between the pattern of roles and rules advocated by standard contracts imposing hierarchical structures on the process and behaviour observed in practice. Building projects are successfully completed through federative mechanisms such as co-operation supported by personal communication and exchange of qualitative and uncertain information. The focus on IT applications is considered to be broadened from the control of cost and schedule, to communication supporting human interaction and mutual adjustment. Electronic links were suggested, with capabilities such as interactivity, simultaneous two-way information exchange and flexibility of communication format, to network all project actors, despite their geographic dispersion and time limitations. Development of a joint IT strategy to enhance a high level of communication in partnering projects is supported by Bennett & Jayes (1998).

Models to shape and describe the concept and various conditions in relationships are developed for partnering. A *partnering continuum* has been created to illustrate the benefit of different degrees of objective alignment between the parties involved. Four stages, each representing a new level of alignment with different applications, are described and illustrated. The traditional approach is called competition and is followed by co-operation, collaboration and, finally, coalescence, representing the ultimate stage of partnering, including a total redesign of work processes between parties. Companies involved in multiple relationships could use the continuum in order to provide guidelines for selecting partners and determining the type of relationship desired (Thompson & Sanders 1998). Another conceptual model was created focusing on organisational boundaries used to classify and describe interfaces between entities as flexible and/or permeable. The model enhances the overall efficiency of the partnering technique by defining the objectives, segregating the resources and establishing a territory for the actors through the modification and addition of new boundaries (Crowley & Karim 1995).

No peer-reviewed papers have been found on the special partnering-leader roles – facilitators and champions. They are described in most of the practical guides and in an informative way in some articles. Champions are involved in the selection and monitoring of facilitators and represent each organisation taking part in large projects. Champions must have a strong link to the next level of command (ECI 1997). They also have a central role in helping to nurture and implement the partnering process. They may well be crucial in promoting and distributing an organisational memory of the lessons learned from partnering experiences (Barlow 2000). Facilitators should be independent, have a basic understanding of construction and knowledge of the partnering process. They should be skilled in organisation, communication, problem solving, conflict-management and listening. Flexibility, willingness to become familiar with the project and the people, and accessibility throughout the duration of the project are important attributes too (ECI 1997).

In an empirical study of nine cases, Bresnen & Marshall (2000c) note that it may still be possible to *engineer* collaboration in the short term, using formal mechanisms such as incentives and teambuilding. These strategies are most likely to be successful where clients already have appropriate experience and capabilities.

Critical views on partnering in construction

Partnering has yet to mature in construction, which is evident in the diversified nature and scope of studies that have been undertaken to date. For that reason, myriad definitions exist (Li *et al.* 2000). The lack of an adequate and precise definition of partnering seems to be a problem. Bresnen & Marshall (2000a) question if it is possible to define partnering as a coherent strategy that involves the deployment of a more or less universal set of practices, systems and procedures. Alternatively, they question why the term partnering is so diffuse and malleable that it can be ascribed to any form of non-adversarial relationship.

Much partnering literature can be characterised as prescriptive. Empirical evidence has been piecemeal and anecdotal with an absence of counter-arguments (Bresnen & Marshall 2000a; Li *et al.* 2000; Green 1999). The way in which partnering has been conceptualised and investigated is criticised by Bresnen & Marshall (2000a), who want to see more pluralistic approaches and explorative studies of the social and psychological aspects of partnering as a mode of organising. Kumaraswamy & Matthews (2000) criticise the fact that the impact of partnering on main contractor/subcontractor relationships has largely been overlooked. Bresnen & Marshall (2000a) support the critique: there is an obvious lack of diffusion of collaborative norms down the supply chain.

Green (1999) views partnering as a crude exercise of buying power and draws the parallel with previous corporatist regimes where different interest groups are bound together by duty and mutual obligations, and where interest groups triumph over individuals. The background to the critique is the gap between the rhetoric of major clients in the UK construction sector and their behaviour. The emphasis on continuous cost improvement makes the logic unsustainable. Continuous improvements, in general, are associated with management-by-stress according to Green. Construction firms cannot be critical because of the threat of being labelled adversarial, thereby denying themselves access to a significant part of the market.

In a systematic critique of previously published partnering literature, academics from China and Australia provide suggestions for future research. Of particular importance are: validation of identified performance measures and critical success factors; development and testing of partnering models and processes; and formatting and selecting partnering strategies. To explain the partnering phenomena and test existing theories there is a need for careful qualitative and quantitative empirical research (Li *et al.* 2000).

The selection process, role of contracts and external environment

Commentators often argue for the contract as a critical component of integration efforts in construction projects (Puddicombe 1997). On the other hand, the argument is that too deep levels of contracting tend to reduce flexibility and the freedom to do what is best to meet the project goals. However, relying on a formal contract alone is not seen as sufficient to promote deeper desired changes in attitude. Behaviour is not determined simply by formal structures and systems, but instead is the result of conscious choices and actions and a complex interplay between structural imperatives and their subjective interpretation and enactment (Bresnen & Marshall 2000a).

Lazar (1997) identified intrusions from the outside world as an external form of barrier to partnering, one example of which is politics. Externally imposed barriers, that affect climate, culture and the structure of organisations, are, for example, laws permitting very strict notice provisions. Another example of an external barrier is a reward system for middle management including disincentives for decision-making and risk taking. Remedies for externally generated barriers to partnering need to be developed on a case-by-case basis.

In creating relationships and contracts where every party should feel like a winner, *equity* becomes a central prerequisite. One *pillar of partnering* is equity, to ensure everybody is rewarded on the basis of fair prices and fair profits (Bennett & Jayes 1998). Short-term commercial sense is one level of equity: another is the more long-term aspect, which can be a question of attitudes to the kind of rewards that partnering can provide. The concept of equity in social transactions can be traced back to Adams (1965), according to whom a condition of equity exists when the values of outcomes and inputs to each party are perceived as proportional. External major changes impact parties' perceptions of efficiency and equity. In terms of contract, the partners' assessments cause them to either engage in renegotiations or to modify their behaviour unilaterally in an attempt to restore balance in the relationship (Arino *et al.* 1998).

Inter-organisational collaboration

The term *alliances* now serves as an umbrella for a host of co-operative relationships in the increasing amount of publications on co-operative activities.

Numerous terms in management are used to describe an alliance, e.g. partnering, partnership, strategic partnership network, co-operative partnership integration, strategic alliance and vertical integration. This phenomenon has drawn a lot of attention from academics during a relatively short period.

Faulkner & De Rond (2000) structure the most popular theoretical frameworks used in empirical studies of co-operative behaviour. Among the economic viewpoints the most important are:

(1) strategic management theory, especially market power theory;
(2) transaction cost theory;
(3) the resource-based view;
(4) agency theory;
(5) game theory; and
(6) real options theory.

Within the organisation field the following are listed as theoretical instruments to treat co-operative strategy:

(1) resource dependence theory;
(2) organisational learning;
(3) social network theory;
(4) the ecosystems view; and
(5) structuralist perspectives.

There is an imbalance in studies, with more focus on the rationale for co-operating, partner selection and performance and less on processes and evolution.

The behavioural aspects are added to the structure as three key areas:

(1) differing culture and the management behaviour they give rise to;
(2) the quality of trust that is so important for all joint endeavour; and
(3) the nature of commitment to the alliance made by the partners.

Partnering in relation to group processes and organisational communication

Partnering is in many aspects a group phenomenon. Within social psychology our knowledge about group processes leads into aspects of team spirit, inter-group co-operation, group productivity, commitment, interdependence and collective problem solving (Brown 2000). By drawing on the strengths of some of the theoretical models in this field, we can try to understand a little more of how to co-operate more effectively in construction projects. The awareness of verbal and behavioural communication skills is another important area of knowledge for the development of effective communication in construction. Organisational communication deals with conflict, persuasion, ethics, roles, rules, culture, networks, diversity, leadership, creativity and technology that all become actual for inter-organisational collaboration in construction (Yuhas Byers 1997).

Research project

Project description and objectives

The research project in its entirety is an empirical replication study with a broad approach to partnering, having the overall purpose of exploring the effects when the concept is used in practice in Sweden. As indicated by Bresnen & Marshall (2000a), there is a need to be cautious when attempting to extrapolate experiences of partnering to other national contexts. In the research, different grounds for using partnering in construction are analysed and a comparative discussion and verification of previous work in the field is being conducted. Some of the conditions that encourage or inhibit partnering in practice will be identified in the context of the cases under investigation. For the reason that all three cases are ongoing, the research results in this chapter are tentative. The focus in the chapter is on the selection process, contract conditions and external environment. The outline of the project is presented in Fig. 14.1.

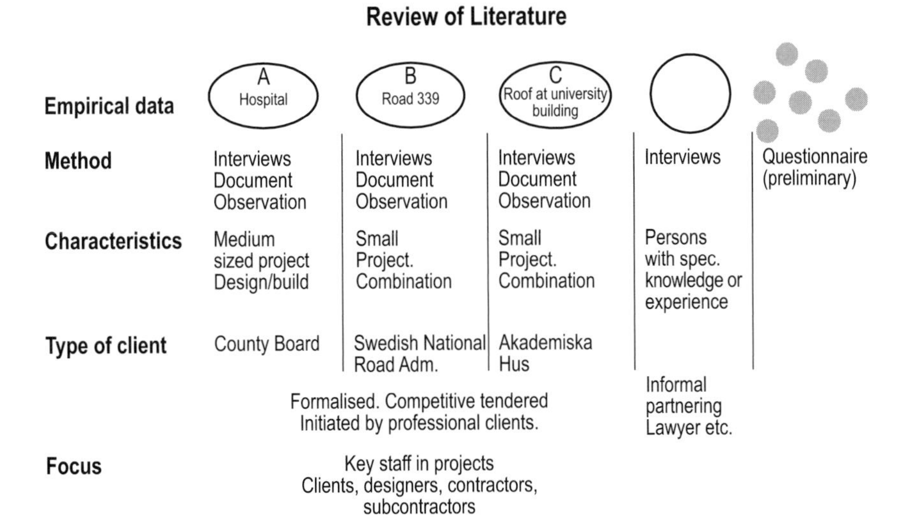

Fig. 14.1 Design of the research project.

In terms of economic value, the projects range from €2.5 million to €34 million and are being undertaken in relatively small local markets. Project A (a hospital) involves serial contracting within a main contract; the others represent small project-specific partnering projects. The period of the relationships is two-and-a-half years for project A, and one year for both project B (a road) and project C (a roof). The price factor in the competitive tendering process was 3% (A), 33% (B) and 40% (C). A target cost including a risk/reward element was used in all projects. There is a mix of approaches to subcontracting between the three cases, but at least the two or three main subcontractors in each case were contracted under partnering arrangements. A document with common goals, known as the charter, was established at a formal workshop in the early stages of the projects. The charters were given high status among the contractual documents.

Design and construction integration included continuous contractor input to design in all cases, with some variations in degree and timing. As a basis for collaboration, two different forms of traditional standard contracts were used. The design/build alternative (A) follows traditional ideas of responsibility strictly in line with the standard contract. Responsibility by solidarity in decisions is, for the combined projects (B and C), included as a condition of contract. Follow-up and induction workshops were present in all projects. In order to enhance communication and the handling of project documents, a project network was used on all the projects. External facilitators with special knowledge of partnering were not present on the projects except B, which was a demonstration of partnering supported by The Civil Engineering Construction Forum, a part of The Royal Academy of Engineering Sciences, *IVA*.

Research methodology

Quantitative approaches have a distinctive function in partnering research for probing patterns and commonalties (statistical generalisations). However, a qualitative approach is used in the study for trying to explore the nature of problems and draw inferences from the data provided. Flexibility and options for in-depth analysis are other reasons for choosing qualitative research methods.

Project-specific partnering concerns dominate over more long-term issues in the study even though the boundaries are not obvious. Case studies are used because the contextual conditions in construction projects are highly related to partnering as a phenomenon. In a case study, it is possible to deal with different sources of evidence and many variables of interest; and furthermore, the form of inquiry does not depend solely on ethnographic or participant-observer data – this was a more pragmatic reason.

A multiple-case design was used with variations within the cases as a strategy. One important practical selection criterion was the availability of formalised partnering projects. Cases were also selected so that a study in real time would be achievable to some extent in all cases. Prior to June 2001, thirty interviews with key staff from different projects had been conducted, with direct participant observation and studies of documentation in each case to supplement the data collection.

The interviews were semi-structured and lasted from one to three hours. In trying to create a holistic picture of partnering in context, the viewpoints of clients, contractors, designers and subcontractors were examined.

Research results and industrial impact

Quantification of results

Competition on price and establishment of the target price, when the scope of the project is relatively undefined, seem to be problematic for contractors. Competitive tendering resulted in a traditional approach to the compilation of the tender for several contractors, and forces related to market power become more apparent in this stage of the process.

Evaluation of tenders on criteria other than price occurred in all cases to a considerable degree. Contractors are unaccustomed to being evaluated on criteria based on opinions such as ability to collaborate, organisation, understanding of partnering and personal skills. After letting contracts, clients have to spend time explaining the motives for their choice to those who did not get the job.

There is evidence of the need for clarity in economic conditions for contracting as, for example, how and when a target price will be upgraded. Every party has to know and understand the share of its own risks before entering into the contract, otherwise the ambiguity will make an impact on the relationship throughout the whole project.

Common goals have influenced to what team members have given priority in the different cases, and these can be in conflict with each other or lead to unexpected consequences. Analysis becomes important for goal formulation work in the early stages.

One example where the impact of the external environment has been extremely high is represented among the cases. Difficulties in reaching decisions at the political level influenced commitment at many other levels in different organisations. This was due to uncertainty in terms of future investment and disturbances in the information flow. On the other hand, collaboration in the form of partnering made it possible to handle major changes in a flexible manner.

A critical factor in developing trust in relationships is time. Another factor from the external environment has to do with the degree of dependency. In a small local market, the parties need good relationships for future work, many of the individuals involved knowing each other from several years of collaboration and who will no doubt meet many times on coming projects. Relationships in these studies are long term after a fashion, even though partnering is for a single project.

Implementation and exploitation

Partnering appears from the outside to be an easy, kind and cosy way of working. Some of those who really have tried partnering think it demands much more of each party than traditional approaches. Habits, attitudes and behaviour being put to the test in difficult or sensitive issues make partnering difficult to achieve. Separately, none of the tools and techniques is new. The challenge is to put these practices together in an effective work setting combined with an awareness of underlying behavioural factors.

The projects studied use some existing tools, but the potential in partnering is not totally utilised. In a national context, the research results can be used as a documentation of early partnering experiences. Strategies for change towards more effective collaboration by the companies involved can be developed from the research, which overall aims to produce additional understanding of the application of partnering in the construction sector. Some new insights may arise, based on the different conditions the Swedish market offers when compared with the UK, US and Australia. The scale of projects, traditions and culture are examples of conditions that might vary between countries.

Conclusions

Partnering is one of the more recent initiatives to improve communication between different parties in the construction process. The problem of insufficient communication has been known for at least four decades, reasons why these problems still remain being many. One could be that the incentive for change has not been strong enough. However, there was almost total agreement among interviewees for a move towards a higher degree of co-operation between parties.

One internationally-known factor influencing the spread of partnering in the construction sector is clients' and contractors' awareness of partnering. This holds good for Sweden too. Deep-seated patterns of behaviour, routines and familiarity with traditional approaches to purchasing, tendering, formulating and evaluating contracts inhibit change; all this in combination with a lack of genuine trust. Change creates uncertainty and can lead parties to reject new thinking and practices. Improvements in relatively small increments are easier to achieve and succeed over time in drawing attention to partnering in small packages. Many clients, designers, contractors, subcontractors and other suppliers need regular practice, and under different conditions, in order to have a chance to establish co-operative behaviour and develop procedures for partnering activities.

Many researchers and practitioners have said that partnering is most appropriate for large projects or projects with a certain degree of complexity. In a smaller project, cost savings or time reduction may be limited. However, there is still a range of other potentially positive outcomes with partnering such as learning, improved quality, end-user satisfaction, safety, cost control and improved working environment. When the project is large and complex, people are prepared to use the concept fully. The conditions for partnering, as a force for change, will be enhanced and the risk of losing another initiative for improving communication is much reduced.

References

Adams, J.S. (1965) Inequality in social exchange. In: L. Berkowitz (ed.), *Advances in Experimental Social Psychology*, **2**, New York: Academic Press.

Andersson, N. & Borgbrant, J. (1998) *Hyreskostnad, förvaltning och produktion i harmoni*, Luleå: Luleå University of Technology (in Swedish).

Arino, A. & De La Torre, J. (1998) Learning from failure: Towards an Evolutionary Model of Collaborative Ventures, *Organisation Science*, **9**, pp. 306-25.

Barlow, J. (2000) Innovation and learning in complex offshore construction projects, *Research Policy*, **29**, pp. 973-89.

Barlow, J., Cohen, M., Jashapara, A. & Simpson, Y. (1997) *Towards Positive Partnering: Revealing the Realities in the Construction Industry*, Bristol: Policy Press.

Bennett, J. & Jayes, S. (1995) *Trusting the team*, Reading: The Reading Construction Forum.

Bennett, J. & Jayes, S. (1998) *The Seven Pillars of Partnering*, London: Thomas Telford.

Bresnen, M. & Marshall, N. (2000a) Partnering in construction: a critical review of issues, problems and dilemmas, *Construction Management and Economics*, **18**, pp. 229-37.

Bresnen, M. & Marshall, N. (2000b) Motivation, commitment and the use of incentives in partnerships and alliances, *Construction Management and Economics*, **18**, pp. 587-98.

Bresnen, M. & Marshall, N. (2000c) Building partnerships: case studies of client-contractor collaboration in the UK construction industry, *Construction Management and Economics*, **18**, pp. 819-32.

Brown, R. (2000) *Group Processes*, 2nd edition, Oxford: Blackwell Publishers.

Crowley, L. & Karim, A. (1995) Conceptual Model of Partnering, *Management in Engineering*, **11**, *5*, pp. 33-39.

DeVilbiss, C.E. & Leonard, P. (2000) Partnering is the Foundation of a Learning Organization, *Management in Engineering*, July/August, pp. 47-57.

European Construction Institute (1997) Partnering in the Public Sector – A Toolkit for the Implementation of Post Award, Project Specific Partnering on Construction Projects, Loughborough: ECI Loughborough University.

Faulkner, D. & De Rond, M. (eds) (2000) *Cooperative Strategy, Economic, Business, and Organisational Issues*, New York: Oxford University Press.

Green, S.D. (1999) The propaganda of corporatism? *Construction Procurement*, **5**, 2, pp. 177-85.

Kadefors, A. (1999) Contractor variation negotiations: a discussion of driving forces and dispositions affecting decision making, *Proceedings of the Nordic Seminar on Construction Economics and Organization*, Gothenburg: Chalmers University of Technology, 12-13 April, pp. 211-18.

Kumaraswamy, M.M. & Matthews, J.D. (2000) Improved Subcontractor Selection Employing Partnering Selection, *Management in Engineering*, May/June, pp. 47-57.

Larsson, E. (1995) Project Partnering: Results of study of 280 Construction Projects, *Management in Engineering*, March/April, pp. 30-35.

Lazar, F. (1997) Partnering: New Benefits from Peering Inside the Black Box. *Management in Engineering*, November/December, pp. 75-83.

Lazar, F. (2000) Project Partnering: Improving the likelihood of Win/Win Outcomes, *Management in Engineering*, March/April, pp. 71-83.

Li, H., Cheng, E.W.L. & Love, P.E.D. (2000) Partnering Research in Construction, *Engineering, Construction and Architectural Management*, **1**, *7*, pp. 76-92.

Liu, A. & Fellows, R. (2001) An Eastern Perspective on Partnering, *Engineering, Construction and Architectural Management*, **1**, *8*, pp. 9-19.

Persson, M. (1999) *Ny Byggprocess-Svedalamodellen*, Lund: Lund Institute of Technology (in Swedish).
Pietroforte, R. (1997) Communication and governance in the building process, *Construction Management and Economics*, **15**, pp. 71-82.
Pocock, J., Liu, L. & Kim, M. (1997) Impact of management approach on project interaction and performance, *Construction Engineering and Management*, **123**, 4, pp. 411-18.
Puddicombe, M.S. (1997) Designers and Contractors: Impediments to Integration, *Construction Engineering and Management*, **123**, 3, pp. 245-52.
Thompson, P. & Sanders, R. (1998) Partnering Continuum, *Management in Engineering*, **14**, pp. 73-78.
Williamson, O.E. (1975) *Markets and Hierarchies: Analysis and Anti-trust Implications*, New York: Free Press.
Yuhas Byers, P. (1997) *Organizational Communication: Theory and Behaviour*, Needham Heights, MA: Allyn and Bacon.

Chapter 15

Importance of the Project Team to the Creation of Learning Within and Between Construction Projects

Fredrik Anheim

Introduction

In recent decades, the rate of cost increases in the construction sector has exceeded the Consumer Price Index. This could be the result of several factors, such as government subsidies, tougher controls, limited international competition, local purchasing behaviour and the oligopolistic situation that exists in several segments of the building materials market. Between 1965 and 1996, productivity in the construction sector corresponded to an average annual increase of 1.7%, compared with 2.9% for other sectors of manufacturing industry (BKD 2000). A feature shared by all successful companies in these other sectors is a desire to learn from mistakes in order to enhance the efficiency of their production operations and be able to offer their customers improved products (Womack *et al.* 1990).

The greater difficulties, when compared with other industrial sectors, that the construction sector has encountered in its efforts to reduce costs, could derive from several different factors. One such factor could be that companies in the sector have found it difficult to learn from the operations that they perform. Each new construction project is viewed as a separate assignment and insight into what could be learned from earlier construction projects is low. Various researchers agree that improved learning could generate highly favourable effects on productivity and quality in the sector (Fernström 1992; Borgbrant 1993; Josephson 1994).

According to Argyris & Schön (1996), an organisation's success is the result of its ability to see things in a new light, to assimilate new realisations and to create new patterns of behaviour. Penrose (1995) is of the opinion that an organisation's growth and development are the result of a collective increase in knowledge within the organisation. If the construction sector could show only a fraction of the pace of development reported for other manufacturing industries, it would be possible to achieve the desired increase in efficiency. Several researchers – for example Senge (1990), Nonaka & Takeuchi (1995),

Ellström (1995), Borgbrant (1987) and Bion (1961) – highlight the importance of the project team in efforts to generate learning in an organisation. Accordingly, the purpose of this chapter is to enhance knowledge of how project teams in a contracting company can affect the potential for its members to learn from the experiences gained in each specific project. This results in the following research question: *how do the project teams of building contractors support or inhibit their members' ability to learn from experience?*

State-of-the-art review

Knowledge and learning – concepts and definitions

Such words as knowledge, experience, competency and learning were assigned a more prominent position in management research during the 1990s. Concepts such as *learning organisations* and *knowledge management* became buzzwords and were claimed to be success factors for organisations in the new millennium (Argyris 1992; Nonaka & Takeuchi 1995; Senge 1990). But what is knowledge? The Greek philosopher Plato introduced the concept of justified true belief. Plato believed that knowledge existed only in our thoughts. Aristotle developed Plato's thoughts that knowledge existed only in the world of thought to include the idea that observations could also provide knowledge. In modern times, Polanyi's (1958) definition of knowledge is significant. He divided knowledge into *tacit* and *explicit* knowledge.

Tacit knowledge can be said to be the knowledge that we as individuals possess without being able to explain what it is to other people, either in writing or in words. It consists of experience and know-how. Accordingly, it is difficult to disseminate tacit knowledge in an organisation. An individual cannot transfer tacit knowledge to someone else, unless the people concerned work together to enable the recipient to experience the knowledge him/herself.

Explicit knowledge is knowledge that can be described in a manner that other people can understand. This means that explicit knowledge can be disseminated in organisations and thus increase an organisation's total expertise (Nonaka & Takeuchi 1995).

Argyris (1993) and Argyris & Schön (1996) claim that learning is something that occurs when we study and correct faults or mistakes, or when we compare results with objectives. They also emphasise the fact that learning occurs in two different stages. The first is when we correct processes or products based on feedback (single loop) and the second (double loop) is when this feedback results in our questioning the values represented by the organisation and any other relevant external sources (Argyris & Schön 1996). In their opinion, learning is also connected to action. They state that when a new action is conducted by individuals or a company, this is proof that learning has really occurred.

Ellström *et al.* (1996) have defined learning as

> '... relatively permanent changes in an individual's competency that result from the individual's interplay with his/her surroundings'.

Björkegren (1999) states that learning is a process in which knowledge is transferred and utilised. Ellström *et al.* also claim that learning involves the breaking of existing routines. If someone possesses new knowledge without changing or reviewing their behaviour – using their knowledge – they have not learned anything. Both individuals and organisations can learn.

For an organisation to be able to learn, it is essential that the individuals in the organisation learn. Learning occurs in the minds of individuals. But if these individuals use their knowledge – their learning – in order to develop the organisation, it could be said that the organisation also learns. Organisations may be viewed as a composition of individuals that possess knowledge. But the organisation itself can also possess knowledge. Examples of such knowledge include procedures for how a specific aspect of work should be conducted, as well as the culture and values existing in an organisation (Argyris & Schön 1996). If the organisation's procedures, culture or values change, it could be said that the organisation has learned: it demonstrates new behaviour.

Thus, learning may be defined as behaviour being changed, or re-assessed as a result of new knowledge – see Fig. 15.1. This definition of the learning process means that the behaviour of the individual or the organisation has been affected by experience to such an extent that the previous behaviour has been re-assessed or changed. The learning process is the process that results in the received experience being registered, converted, communicated, understood and used.

According to Senge (1990), a learning organisation could be said to be a company that utilises its employees' commitment and ability to learn at all levels of the organisation. The aim is for the company to develop faster than its competitors. Nonaka & Takeuchi (1995) focus primarily on what they call *knowledge creation*. They believe that new knowledge is created when people with different experiences (tacit knowledge) meet with the aim of solving different problems. New knowledge is created when this meeting of the different individuals' prior knowledge and experience is combined with a challenging mission.

Fig. 15.1 The learning process.

Knowledge transfer – concepts and definitions

Knowledge can be transferred in various ways. It can be transferred by an individual describing an experience to someone else, via written instructions, through an illustration of an object, through a metaphor or by looking at how somebody else performs an action. How well (meaning how authentically it is understood by the recipient) such knowledge is transferred depends on several factors. All individuals have different prior knowledge and are affected by different environments – they exist in different cultures. As a result, they interpret the information in different ways (Reddy 1993).

Sveiby (1995) states that knowledge transfer can be divided into two different processes. The first process, which he calls information, occurs indirectly through some form of medium and the second, which he calls tradition, occurs when individuals exchange knowledge directly with each other. Knowledge management can be viewed as a type of knowledge transfer within an organisation and may be defined as how a company utilises the knowledge possessed by the people in the organisation. Mårtensson (1999) says that knowledge management makes visible, and highlights, the competencies available in an organisation with the aim of subsequently being able to control the organisation in a better way.

Björkegren (1999) argues that the transfer of knowledge between different projects in an organisation is subject to obstacles and opportunities. She emphasises two main methods for transferring knowledge between projects. The first occurs through knowledge bearers, meaning individuals who move from one project to the next, bringing with them knowledge and previous experience. The second occurs when the experience results in changed procedures for controlling how work in forthcoming projects is to be conducted.

Team learning

To facilitate organisational learning within a company, various driving forces or other key prerequisites must be discussed. A number of leading researchers in this field state that working in a team is one of the principal prerequisites in this context. A few of the most relevant of these theories are presented below.

Nonaka & Takeuchi (1995) have built on Polanyi's definition of tacit and explicit knowledge by focusing their interest on identifying where in a company this learning occurs. They believe that new knowledge arises in a company when people possessing tacit knowledge meet and gain an opportunity to exchange their experiences with each other, a process they call *socialisation*. The tacit knowledge can then be converted into explicit knowledge that can be assimilated by other people in the organisation. Fischer (1999) developed ideas about how tacit knowledge becomes explicit knowledge. He believes that it is not possible to move directly from tacit knowledge at the individual level to explicit knowledge at the team level. The knowledge must be changed from tacit to explicit knowledge before it can be disseminated in the organisation.

Nonaka & Takeuchi (1995) also place considerable emphasis on the process needed to convert tacit knowledge into explicit knowledge, and what makes the knowledge understood throughout the company. The existence of a dynamic team is regarded as the principal factor in this process. By mixing people with different skills and prior knowledge, tacit knowledge can be understood and made explicit. Working in a team also results in the explicit knowledge being disseminated and understood by an increasing number of people in the organisation. In addition, the interaction that occurs within the team leads to the creation of new knowledge through the combination of the various team members' knowledge. This phenomenon was also noted by Kjellberg (1996) who believes that *target-oriented teams* play a decisive role in determining whether the people in a production system will be able to contribute their knowledge.

Learning organisations are hallmarked by the existence of a new type of leadership and cross-functional co-operation. The existence of cross-functional co-operation in teams throughout the process is the very factor that determines whether or not the competencies of the team members will be utilised and thus whether or not learning will occur. Kjellberg defines *competency* as a person's ability to *act and to utilise his/her skills and knowledge.* To ensure that members of target-oriented teams will be able to utilise their competency, it is essential that they be given the scope for making decisions and be subject to demands that match their knowledge.

Senge (1990) describes five different disciplines that are prerequisites for the creation of a learning organisation. The fourth discipline – team learning – is based on a dialogue conducted in teams. Today, almost all training and development work in companies is conducted in teams. If the collective knowledge represented by these teams can be used to enhance the competency of the entire company, the pace of development will be faster than if each individual member learns the subject independently up to the same level of knowledge. For team learning to function, it is therefore essential that a climate be created within the company that facilitates a free exchange of opinions.

Ellström *et al.* (1996) also emphasise that teamwork is an important arena for exchanging experiences. They also note that working in teams gives rise to both positive and negative consequences. The negative consequences include the existence of conflicting interests and territorial thinking within the team. An *us and them* atmosphere could arise between various personnel categories, thus inhibiting the team's learning capacity. Ellström *et al.* also believe that the team's composition is a significant factor with respect to the ability to function as a learning arena. He states that both an excessively homogeneous and an excessively heterogeneous team composition could have an inhibiting effect.

If the team is too heterogeneous, destructive conflicts could arise. If the team is too homogeneous, an excessively uniform way of thinking would arise, as well as resistance to learning something new. To facilitate the generation of the positive learning effects that a team is expected to provide, its composition should reflect considerable differences in the members' individual competencies, but only small differences in terms of values and interests.

Borgbrant (1987) argues that certain criteria are of key significance if a project team is to function well, namely participation, good work environment, rewards for contributions and authority versus responsibility. Hägerfors (1995) notes that individual members of a team need to feel affinity to the other team members, to feel security within the team and to respect the other members if the team is to be effective. Bion (1961) points to the relevance of the team as a possible development environment for the individual team members. He highlights seven factors of key importance to the creation of a healthy development environment: shared goals, shared values, the ability to accept new members and to adapt the team accordingly, not permitting subgroups with their own delineations, freedom for individual team members, an ability to resolve conflicts within the team and ensuring that the team consists of at least three members.

Another factor highlighted by both Senge (1990) and Nonaka & Takeuchi (1995) is the importance of shared goals. If people do not know where they are headed, it is difficult for them to understand what aspects of operations are important. Bion (1961) states that all teams that meet, regardless of the purpose, have a goal for meeting. He believes that it is this goal that results in the team co-operating to resolve a task. Ellström *et al.* (1996) also assert the importance of shared goals in efforts to motivate individuals to learn. They believe that the goals are often vague and, most importantly, that they often change during the course of a project. They also claim that ensuring that the individuals in a team participate in the goal-establishment process is at least equally important. They must be allowed to take part in formulating the goals, and to reflect on and reformulate the goals that have already been established for the business operation. Borgbrant (1987) also notes this factor. He believes that it is essential that team members feel an affinity with the organisation's goals and that they be allowed to participate in following up how achievements correspond with the goals set.

Research project

Project description and objectives

Due to the structure that is applied to the execution of construction projects, considerable amounts of information and knowledge are lost. Many aspects of the construction process are performed by different actors that have no contact, or only limited contact, with each other. The players that enter the process at a late stage have limited opportunities to affect the solutions that have been selected. This is due, in part, to the difficulty of making changes in one area because changes also affect other parts of the project and, in part, due to contracts having been signed for a specific type of project design. The strict division of functions among different contractors also has an adverse impact on the potential for learning. There is a division between those who work indoors in offices and those who work outdoors at the construction site. This results in an *us and them* atmosphere that inhibits effective co-operation. Such an *us and them* atmosphere is at least equally apparent between skilled workers and salaried employees involved in a construction project. The duties of each category are abundantly clear.

The commercial design and general planning are conducted by personnel stationed in the permanent office, the detailed development planning and production control are carried out by the administrative employees involved in the construction project, while the craftsmen perform physical production work. Accordingly, several information gaps arise during the production stage. The ideas that the salaried employees had when they calculated the cost of production are not always communicated to the craftsmen when they have to perform their various duties. Often, the craftsmen themselves determine how various work aspects are to be performed, based on personal experience. In many cases, the team foreman's influence on how the work is performed is equal to that of project management. As a result, the ideas that were conceived at one stage – whether they were right or wrong – are not used.

A contracting company should actually have an advantage in terms of learning because natural project teams are formed on each construction project. In the construction sector, there is a long tradition of conducting projects using a team of craftsmen managed by a group of salaried employees. These individuals are on-site and share a common goal, namely the construction of a building. As mentioned in the introduction, however, previous research indicates that this environment does not function as an arena for learning.

In this project, the type of research conducted is mainly that of evaluative design. The aim of the research is to study the building contractors' working methods and how teamwork affects the potential for learning. To a certain extent, the research is also change-oriented, since its purpose is to study what must be changed in order to effect an improvement in the learning that occurs within the organisation (Andersson & Borgbrant 1998).

Research methodology

In terms of method, this work may be described as inductive. During the entire research process, the researcher has sought to find deeper insight into the matters that have proven to be of decisive importance to achieving the aims of the project. A number of assumptions made in previous research are named in the problem description. No further studies of these assumptions have been made. The research is based on qualitative analyses of the data collected. Qualitative analyses were selected because of the nature of research involved (Jensen 1995). In order to understand the driving forces with the potential to generate learning, deeper understanding of the individuals involved would be required. It is very difficult to achieve such understanding in quantitative studies. Since this research is not based on hypotheses, the researcher does not know which factors are of importance to answering the research questions asked at the beginning of the project. Accordingly, a qualitative analysis of the data collected could result in knowledge that indicates new, unexpected relationships. In areas considered appropriate, however, quantitative studies have been used in order to broaden the collection of data.

In qualitative analyses, considerable demands are placed on the researchers who themselves constitute a significant instrument in analyses of the results. The individual researcher's scientific approach, and experience of life, influence the way the collected data are analysed.

The researcher is employed by the company being studied. This naturally gives rise to difficulties in terms of the need to maintain distance and objectivity when analysing collected material. It is, of course, difficult to determine how successfully this has been done. However, the fact that the researcher has been conscious of this throughout the study hopefully means that this factor had only a limited effect on the survey's reliability and validity. The fact that the researcher is employed by the company under investigation can also yield favourable effects (Andersson 1979), since being employed can enable the researcher to make closer contact with the respondents and blend in more naturally during the execution of the case studies.

Research results and industrial impact

Quantification of results

The study upon which this chapter is based was conducted on two different construction projects. In addition, a benchmarking study was made of a large industrial company. With respect to the case studies, the members of project teams were asked to respond to questions showing their opinions of how well co-operation within the project and between other projects had functioned. Although several interesting aspects arose during the study, this chapter focuses solely on the importance of the team to the potential for learning within a project and between projects.

The theory noted in the preceding chapter sheds light on several factors of significance to the achievement of efficient learning in teams. The factors have been used as a screen against which the material gained from interviews has been reflected. A number of factors have been selected for the analysis of collected data – see Fig. 15.2.

In the case study, seven factors were selected as being significant for *team learning*, namely: *arena for learning, tacit to tacit, combination, high tolerance, scope for decisions, composition* and *dialogue* (Fig. 15.2).

The potential to convert tacit into explicit knowledge is considered limited. The types of meetings in which individuals sit down to exchange experiences and reflect on their importance have not been identified in the case study. Cross-functional groups including people outside the project also appear to be unusual. Within the project, there did not appear to be much co-operation outside the various subgroups.

A number of the seven identified factors warrant additional comment. The opportunity to combine different types of knowledge exists but would function much better if cross-functional contacts were established, and if the membership turnover within the various teams was greater. It is also questionable whether the teams are actually characterised by high tolerance. To facilitate learning and exchanges of experiences, the climate should be such that individuals are not afraid to make mistakes. Construction projects are often hallmarked by a *harsh yet hearty* climate. The question is: how does this affect the individuals' willingness to admit mistakes?

The team's composition was identified as a positive factor in the case study. However, several respondents stated that project teams are not usually composed in the same manner as the construction project covered by the case study. The projects usually have a more homogeneous team composition with members who accompany each other from project to project. Dialogue does occur, but only within an individual subgroup. The occurrence of dialogue with other subgroups within a project and with other projects is very limited.

Team
Arena for learning
Tacit–tacit
Tacit–expressed
Combination
Cross-functional
High tolerance
Scope for decisions
Composition
Shared goals
Dialogue

Fig. 15.2 Factors affecting team learning.

The construction sector has a strong tradition of working in projects and of organising employees in teams that accompany each other from one project to the next – this applies to both craftsmen and salaried employees. The interviewees were of the opinion that they worked in teams during the construction project. They believed that a lot of effort was devoted to measures aimed at enhancing job satisfaction within the construction project, in order to generate a feeling of affinity among all the parties involved – meaning both the company's own employees and those of subcontractors. It would be very difficult to complete a construction project if work was not conducted in teams.

According to the data collected, the employees believed that they worked well in a project structure and that this type of teamwork was essential if all of the duties of a building contractor were to be conducted satisfactorily. In other words, an *arena for learning* exists. However, there were several subgroups in this case study: salaried employees, the team of construction workers, the various installation subcontractors, project management and so forth. The transfer of *tacit knowledge* seemed to work well in all of the subgroups. Craftsmen apply an apprenticeship system based on the idea that tacit knowledge is transferred when the apprentices see how the master performs a task and then try to perform the task themselves. Converting tacit knowledge into explicit knowledge appears to be more difficult.

Conclusions

To a large extent, learning in a construction project takes place at the level of the individual. Individuals identify various events within a project and then make their own reflections regarding the importance of these events. A certain amount of learning also occurs in the subgroups that exist within projects. When an important issue arises, the matter is discussed within the subgroup, which results in joint team learning.

The study also indicates that the teams existing in construction projects contain certain features that are positive with respect to learning. Projects are autonomous since considerable freedom of action exists within the framework of a project. The construction projects studied involved the production of a complex product that generated many learning opportunities. Nonaka & Takeuchi (1995), Björkegren (1999) and Ellström *et al.* (1996) all highlight the significance of a complex learning environment, stating that a sufficiently stimulating task is a prerequisite for learning. Construction contracts may be viewed as complex projects that contain a large number of different problems.

The case study mainly points to three deficiencies in the construction company's teamwork. It appears that project teams are kept excessively intact from one project to the next; there are several subgroups within the projects; and the individuals in the subgroups have insufficient contact with other individuals and with groups outside the project.

Team composition is an interesting factor. According to Ellström *et al.* (1996), there is a risk that teams that work together too long will become overly homogeneous. Such teams conduct all of their work as a matter of routine. He emphasises the need to bring in new impulses to a team. If a team does not change, there is a risk that it will not function as a learning environment. Ellström *et al.* (1996) and Bion (1961) point to the fact that subgroups with different interests can develop an *us and them* atmosphere that can give rise to an adverse impact on the team's capacity to function as a good learning environment.

References

Andersson, N. & Borgbrant, J. (1998) *Byggforskning-Processer och Vetenskaplighet*, Luleå: Luleå University of Technology (in Swedish).

Andersson, S. (1979) *Positivism kontra Hermeneutik*, Gothenburg: Bokförlaget Korpen (in Swedish).

Argyris, C. (1992) *On Organizational Learning*, Oxford: Blackwell Publishers.

Argyris, C. (1993) *Knowledge for Action*, San Francisco: Jossey-Bass Publishers.

Argyris, C. & Schön, D.A. (1996) *Organizational Learning II*, Reading, MA: Addison-Wesley Publishing.

Bion, W.R. (1961) *Experiences in Groups*, London: Tavistock Publications Limited.

Björkegren, C. (1999) *Learning for the Next Project*, Licentiate thesis, Linköping: Linköping University.

BKD, Byggkostnadsdelegationen, Swedish Delegation on Building Costs (2000) *Från Byggsekt till Byggsektor*, SOU 2000:44 (in Swedish).

Borgbrant, J. (1987) *Strategisk Dialog*, Stockholm: Natur och Kultur (in Swedish).

Borgbrant, J. (1990) *Strategisk Dialog 2*, Stockholm: Natur och Kultur (in Swedish).

Borgbrant, J. (1993) Lärande som framgångsfaktor, in Engfors C. (ed.), *Lära om Hus*, Stockholm: Arkitekturmuseet, pp. 66–71 (in Swedish).

Ellström, P.-E., Gustavsson, B. & Larsson, S. (1996) *Livslångt Lärande*, Lund: Lund University (in Swedish).

Fernström, G. (1992) *Byggbranschen på Nittiotale*, Stockholm: Byggförlaget (in Swedish).

Fischer, W.A. (1999) NCC R&D Conference 1999, Unpublished conference material.

Hägerfors, A. (1995) *Att Samlära i Systemdesign*, Lund: Lund University (in Swedish).

Jensen, M.K. (1995) *Kvalitativa Metoder*, Lund: Lund University (in Swedish).

Josephson, P.-E. (1994) *Orsaker till fel i Byggandet*, Gothenburg: Chalmers University of Technology (in Swedish).

Kjellberg, A. (1996) *Produktionssystemets Pedagogic*, Stockholm: Royal Institute of Technology (in Swedish).

Mårtensson, M. (1999) Knowledge management – kunskapsarkivering eller kunskapsaktivering? in *Knowledge Management – Kunskapsarkivering eller Kunskapsaktivering? – årsbok 1999*, Stockholm: Sweden's Technical Attachés pp. 13–27 (in Swedish).

Nonaka, I. & Takeuchi, H. (1995) *The Knowledge Creating Company*, Oxford: Oxford University Press.
Penrose, E. (1995) *The Theory of the Growth of the Firm*, Oxford: Oxford University Press.
Polanyi, M. (1958) *Personal Knowledge: Towards a Post-critical Philosophy*, Chicago: University of Chicago Press.
Reddy, M.J. (1993) The conduit metaphor, in Ortony A. (ed.), *Metaphor and Thought*, Cambridge: Cambridge University Press, pp. 164–201.
Senge, P.M. (1990) *The Fifth Discipline: The Art and Practice of the Learning Organization*, New York: Doubleday Publishers.
Sveiby, K.-E. (1995) *Kunskapsflödet*, Borgå: Svenska dagbladets förlags AB (in Swedish).
Womack, J.P., Jones, D.T. & Roos, D. (1990) *The Machine that Changed the World*, New York: Rawson Associates.

Chapter 16
Refurbishment of Commercial Buildings: the Relationship between the Project and its Context

Åsa Engwall

Introduction

In general and at the strategic level, the real estate and construction sector has the following characteristics: long life cycles; long development cycles; high uncertainty regarding customer evaluation; and high uncertainty regarding a specific building in comparison with other buildings. Within this context are tenants, whose needs change over time and that may not be satisfied by their existing premises. One response is for owners to refurbish or redevelop their buildings to discourage tenants from moving to other, more suitable premises.

At the operational level, the process can be described as an organised collection of specifications – either frozen or fluid – interpreted and codified in building performance terms within a given procurement method. The many battles between the main actors – client and building contractor – suggest contradictions concerning the requirements that a building project is intended to meet (Fenn *et al.* 1997). A possible explanation might be that a building's performance undergoes change during the contract period; a change of plan can create a dynamic and continuing activity between an actor and his/her environment. This implies that the planning base is unstable and project uncertainty is probably of particular concern.

Research has so far supported the proposition that performance in new product development processes depends on the organisation's ability to perceive uncertainties regarding the external environment (Duncan 1972; Christensen & Kreiner 1991). However, few studies in the real estate and construction fields report on how organisations engaged in the construction process respond to uncertainty. The objectives in this chapter are, therefore, to examine the conditions within which project work is performed in the construction process. Particular attention is given to problems arising during the process that affect the relationship between the client and the (main) contractor and/or the product. The response of project organisations in terms of how they handle project uncertainty is of particular interest. A case study, based on a client's perspective of the process, is presented. In this case, the client is a real estate company that owns, operates and markets offices in a major building redevelopment.

Theoretical framework

The current practice of dividing the construction process into phases suggests that the process is seen as a logical, planned and disciplined activity. The procurement method assumes, or makes explicit, certain informational requirements, especially those of activities and means. This implies that customer needs should also be explicitly formulated and stated in plans ahead of construction. The definition of the management task is therefore believed to be nothing more than the operational planning of such needs and interests in the construction process (Kreiner 1995). The reality is somewhat different. For this reason, the traditional picture of construction theory has been extended with reference to theories of new product development, theories of uncertainty and theories of product assessment. It is believed that this extension provides a wider perspective for understanding and describing refurbishment processes in the real estate and construction sector.

A fragmented model of planning and construction

In Sweden, the essential concepts of construction planning and production stem mainly from the 1960s (Eriksson 1994; Sandström 1994). At that time *Tayloristic* ideas were implemented as tools to serve a rationalised construction process, with the aim of increasing overall industrial productivity. The model of planning and production was characterised as a sequential, linear process, starting with planning and design activities, progressing to production and ending with the handing over of the completed building for occupation and periodic maintenance.

The same principal concept was taken to govern the construction process regardless of whether it concerned a stable situation, for example, a project for a defined customer, or an experimental situation such as testing new technology. There was a strong pressure to standardise technical, administrative and also organisational management; for example, standard forms of tender procedure and emergent job specialisation. The same process was used even in situations of a speculative nature (i.e. no defined customer). By the mid 1970s, the value to the customer of the working environment came to be seen as a necessary ingredient for success in the planning and design phases (Sandström 1994).

There was a growing understanding of the need to develop customer value and not to treat planning and production processes as isolated activities.

The utility of this rational building model has been hotly debated. Criticism has focused mainly on the fragmentation among participants; the fragmentation of design and construction data; costly design changes; the lack of life cycle analysis and the lack of communication between actors in the process (Evbuomwan & Anumba 1996; Anumba & Evbuomwan 1997; Pietroforte 1997).

New product development

A key issue in product development is how firms actually develop new products. In a literature review, Brown & Eisenhardt (1995) discern three characteristic streams of past research in new product development. These are:

(1) product development as a *rational plan;*
(2) product development as a *communication web;* and
(3) product development as *disciplined* and *integrated problem solving.*

The first discipline favours a *rational plan.* It emphasises that product success is the result of good (i.e. careful) implementation of standard activity plans, significant support from senior management and financial performance analysis: for instance, product specifications, clear product concepts and market share. Overall, this stream reflects standard, well-defined operational plans (procedures or processes), and internal organisation factors as guidelines for development activity.

The second stream, the *communication web,* looks at new product development as a result of communication among project members, fuelled by external stimulation. Overall, this stream examines information and information exchange and the key actors in product development activities. These theories suggest that series of information acquisition and the actual use of information are key aspects of development activities (Galbraith 1973, 1993; Kreiner 1995; Katz 1997; Nadler & Tushman 1997). Key actors in these activities are recognised as individuals who bring information into the organisation and disperse it to fellow team members. Studies that consider the impact of social interaction on innovation recognise communication barriers and communication support as affecting product performance. Acona & Caldwell (1992) show that teams that have more external communication flow (i.e. teams with a lobbying and political approach), and teams that have thorough internal communication (i.e. a cocoon-like existence, ignoring the outside world), perform better than others.

The third stream, *disciplined integrated problem solving*, draws attention to operational performance and decision-making with reference to Japanese custom, as presented by Nonaka & Takeuchi (1995) and Womack *et al.* (1990). The style of product development is described as comprising a cross-functional and relatively autonomous project team, a product vision, powerful leadership, an iterative prototyping process and high supplier involvement. The development process itself is described as a rapid cycle of overlapping activities.

Song & Montoya-Weiss (2001) use a conceptual model based on the rational plan and the disciplined problem-solving concept. They also identify the necessity of a political and a resource approach to explain the project outcomes of product development. Their model specifically recognises that project performance is strongly affected by the product development process. Project performance refers to a product's perceived superiority relative to competitive products and the level of financial success achieved by the new product. Beside the technical perspective, Song & Montoya-Weiss (2001) suggest that models of product development should emerge to include marketing activities and perspectives of competitive and market intelligence.

The impact of uncertainty

Uncertainty and, specifically, task uncertainty were identified above as one of the conditions of product development. Past research suggests that there are various types of perceived uncertainty in the environment; for example *technological uncertainty, consumer uncertainty and resource uncertainty* (Duncan 1972; Jauch & Kraft 1986; Song & Montoya-Weiss 2001). Lawrence & Lorsch (1967) state that the term *uncertainty* consists of three components:

(1) the lack of clarity of information;
(2) the long time span of definitive feedback; and
(3) the general uncertainty of causal relationships.

With regard to construction, Kadefors (1992) and Galbraith (1973, 1993) explain three key issues for determining levels of uncertainty: the number of products, services and customers; the fragmentation of skills; and the level of performance. Duncan (1972) showed that the complexity of the environment increases as the number of components increases and as the components become more unlike each other. In development processes there is also a need to analyse processes or activities that are interdependent. Such activities need co-ordination and sharing of information (Thompson 1967).

Christensen & Kreiner (1991) claim that the organisational dilemma is to understand and to cope with different types of uncertainty. The central concern for an organisation, therefore, is to establish an understanding of conditions for a project and project work. Uncertainty should be understood as being of a dual and coherent nature with reference to the environment, comprising the building context and the management of construction operations for the specific building. Two concepts are therefore established: *contextual uncertainty* and *operational uncertainty.* Christensen & Kreiner (1991) propose that these types of uncertainty are typically present in construction.

Contextual uncertainty includes the environment as a whole that may have an impact on a specific building. The impact might raise doubts about the result or the effectiveness of the achievement. In order to analyse the shape or the forms of contextual uncertainty it is necessary to analyse the building from a broader perspective: the environment, the customers and the organisation as a whole. The character of contextual uncertainty can be determined only after the completion of the project. The issue is to determine *what* existed within the environment. The difference between the environment before and after completion of the project is the definition of the contextual uncertainty for that project.

Operational uncertainty is defined as every circumstance that may have an impact on project efficiency; that is, handling the implementation of construction according to a predetermined set of goals. The logical phases of the construction process mean that project visions are needed to reduce planning and design uncertainty, and that plans are needed to reduce production uncertainty. Actions taken in order to reduce managerial operational uncertainty may have an impact on project achievement. The model sets limits so that the project may be handled in a realistic manner from the start. The model also tends to put formal rules ahead of innovative approaches. Models of rational planning are used to reduce operational management uncertainty. Another trend is to keep the project on the track of its original plan. This model requires that the project is kept isolated from its environment. The risk of this isolation is that the project may run into a sidetrack as a result of events in the environment. The management of operational uncertainty relates to the task of the project team, which is selected to solve problems and to turn plans into reality.

Assessing product value

Schuman *et al.* (1995) offer a framework that can be used to identify explicitly the objectives and the focus of measurements for product evaluation. They claim that there are numerous ways of assessing product performance, such as cost of product processing, patents, external testing and benchmarking. These instruments of analysis reflect different perspectives such as marketing, organisation, design and operations management (Krishnan & Ulrich

2001). Despite all these testing activities, Schuman *et al.* (1995) argue that the customer should determine the best standard value of a product. It is also presumed that a product that matches the requirements and mirrors the values of the customer can generate revenues (Schuman *et al.* 1995; Preece & Male 1997; Holm 2000).

In general, the objectives of a client in construction are stated quite simply (Ward *et al.* 1991). The objectives concentrate mostly on cost, time and quality. Schuman *et al.* (1995) recognise that the client may be a company providing services or products to others. These clients must perform and satisfy their own customers. In such situations the components of customer satisfaction are clearly complex. Among the most noted customer satisfaction components are product capability, utility (fitness), product performance, reliability and ease of installation and maintenance. Assessment of the consequences of a product may also include judgements of cost, timely delivery and service, together with product attributes. Contributions to health and job creation, and social and political impacts, may also be identified as customer product values.

In construction, Liu & Walker (1998) argue that the complexities underlying the evaluation of project outcomes, that are derived from project goals, participants' behaviour and the performance of project organisations, need to be properly understood. Furthermore, Winch *et al.* (1998) claim that construction management texts do not mention the customer, yet advocate an orientation towards delivering customer satisfaction. In his study, Mahmoud-Jouni (2000) concludes that market offerings in construction must become more strategic in order to improve product targets. He suggests that construction should consider (1) product offering and/or (2) technical offering, and that work methods and their relationship must, therefore, be redefined.

The need for a broader perspective

Several writers have recognised uncertainty as one of the dilemmas in the construction sector (Christensen & Kreiner 1991; Ward *et al.* 1991; Kadefors 1992; Mahmoud-Jouni 2000). However, few studies in real estate and construction management report on how organisations in the construction process act and respond to project uncertainty and its relationship with the environment, and how the environment might affect project performance. Furthermore, construction literature often concentrates on procedures, rather than on the need for providing value for the client or the customer (i.e. tenant in this case). This approach is recognised by other writers such as Winch *et al.* (1998), Liu & Walker (1998) and Holm (2000). Clearly, past research mirrors the fragmented construction process, another key dilemma within the sector.

Past research tends to analyse and describe the process from the constructor's viewpoint, while the viewpoint of the client is rarely mentioned. The process of new construction is mentioned, but refurbishment is not. The underlying assumption in the literature is that uncertainty is bad for the organisation and that uncertainty in the construction process can be reduced by well-defined and careful procedures established prior to construction. To match this rational idea of a construction process is the assumption that tenant needs are stable, their perceptions of product success are fixed and that, therefore, it is possible to predetermine needs. This might not be the case. This chapter therefore seeks to contribute to a more complete and balanced project orientation, demonstrated in the following case study of a refurbishment project.

The case study of Oxenstiernan

Background

During the period from mid-1998 to the end of 2000, an office building known as Oxenstiernan, in central Stockholm, was refurbished. The building was planned and constructed in the early 1960s to serve parts of the Swedish military. In the refurbished building, the total provision of rental space was in the region of 21 000 m^2 covering 1 000 individual offices, with a total cost for the refurbishment estimated at SEK 350 million (€39 million) at 2000 prices. An additional 10 000 m^2 of rental space in a connected building was also part of the work.

Method

The research method is essentially a case study concerning the Oxenstiernan office building. Data were collected through interviews with 32 people connected with the project, and from direct observations and studies of project documents. The interview scheme represented the client organisation (24 people), the main contractor (4 people) and tenants (4 people). The interviews were conducted in a semi-structured manner (Robson 1993; Yin 1981, 1993). Most interviews lasted for one-and-a-half hours, but some lasted up to four hours. The purpose of the interviews was to understand people's perceptions of the project outcome (product achievement) and its correlation with the refurbishment work. Since the intention of the study is to provide an understanding of causes and effects in one single case, data for statistical analysis are not presented.

The case study was conducted over a two-year period. This allowed issues to be studied as they arose during production and their connections with planning and design activities, as well as following up project achievements in maintenance and tenant post-occupancy evaluations. Direct observations were performed as walking tours of the construction site and in tenants' offices. The main reason for this was to move from talking about issues to seeing actual outcomes. It was also an opportunity to validate oral statements. Studies of planning documents, drawings and minutes from project meetings were undertaken. These documents typically dealt with financial issues or intended outputs. In general, there was a lack of information on the background to decisions and of discussions on alternative approaches.

The findings were validated in two ways; first, by observed patterns made by Anheim (2001) who studied the same project from the contractor's perspective; and second, in discussions with co-workers from the client firm and the (main) contractor firm at a workshop in autumn 2000.

The client perspective

In general, the objectives of the client can be stated in terms of cost, time and performance. Objectives can also take on marketing and technical perspectives. Marketing proficiency includes activities such as evaluating customers and competitors, determining market trends and market research. Technical proficiency comprises activities that entail production and engineering evaluations, determining product specifications and construction of the final product.

Technical proficiency

Apart from a technical specification to support a modern office environment, another objective was in line with the real estate owner's motivation of achieving consistency between the internal and external metaphor of the office building. This was for IT (information technology) enhanced offices in which building and facilities management were performed properly. Consequently, the tender for the refurbishment project required the contractor to make arrangements for the use of IT, as well as setting up a client-server network during the on-site production.

The ratio between net lettable area and gross floor area was 0.67. The estimated cost of SEK 292 million (€32 million), at 1997 prices, for refurbishment work compared with revenue was considered to be high in this project. The original plans show offices with a modern form of linoleum flooring, preferably for an open and flexible arrangement of space, with a few individual offices. Services included toilets with a form of plastic flooring and kitchenettes (i.e. a simple kitchen with no cooking facility), as standard for this kind of project.

Originally, on-site production was scheduled to last for 18 months from August 1996, but needed rescheduling several times as the issue of the building permit was delayed because of appeals to court and the EU. A decision was made in April 1997 that would have allowed construction to start in July 1997 and enabled tenants to move in from May 1998. However, the building permit had not been issued. Demolition work was allowed to start and eventually, in the spring of 1998, work began on-site. The cost of the building permit delay was estimated at SEK 85 million (€9.4 million) in total at 1997 prices.

The project was based on a traditional contract with a main contractor who had to co-ordinate all the actors in the project and interfaces between the different building systems. That said, the client handled planning and design, as well as the operation of the building.

On-site production work was divided into two parts. The first part covered the entrance areas and communication systems, the structural frame and HVAC. The second part covered the customisation of each office. Realisation of the design was the subject of 26 separate contracts, and the completed refurbishment was estimated to be SEK 350 million (€39 million) at 2000 prices. Out of the SEK 58 million (€6.4 million) excess, SEK 10 million (€1.1 million) was the result of further taxation. Increasing the building's performance resulted in additional costs of SEK 12.2 million (€1.35 million); further customisation of offices added costs of SEK 21.4 million (€2.37 million); and eventually SEK 14.4 million (€1.55 million) was added as a consequence of market inflation. The result of increasing the building's performance was an increase of close to 2% in the project's return.

Marketing proficiency

Before refurbishment started in 1998 the office building was regarded as an unpleasant ghost-house. It was felt that the refurbishment would serve as a prominent example project within the client organisation and therefore stand as a good reference for the organisation's reputation for competence. Throughout project planning, design and production, changes were made successively to improve the performance of the building. A shift in the commercial market became part of the project vision. This vision was essentially a reaction to a significant and growing interest in the financial district located 10 minutes walk from the building.

The refurbishment cost was equivalent to one third of the annual investment budget of the client. The project, therefore, represented financial as well as symbolic values. Of major concern to the client were the entrance of the building, the physical environment and the building permit. During the planning and design phases, the building was found to be four storeys too high according to building regulations. The building originally lacked an entrance, being connected to another building for reasons of military security. Now tenants' and visitors' perceptions of the entrance became important considerations. The aesthetical treatment also affected part of a larger entrance refurbishment of the total existing building, amounting to some 180 000 m^2.

Between 1996 and 1997 ideas were presented in order to test the market for IT-enhanced offices. At that time there was only one office building in Stockholm that had been linked to the emergent e-economy. Exactly what was needed in terms of a building or buildings to support the e-economy was initially unclear. Finally, a decision was made on an IT marketing profile for the building.

The military moved out during 1996, and a first programme for refurbishment was proposed during 1995. At that time the client tried, unsuccessfully, to reach agreement with a large Swedish company that could fill the whole of the building. One tenant of 2 600 m^2 had agreed on rental terms after the refurbishment, but no other agreements had been reached. It was assumed that suitable tenants for the office building would be in the business range of privately-owned, knowledge-creating firms and others with an interest in the business of education. A minimum rental space of 200 m^2 was set and the appropriate number of tenants was estimated as 17. At the start of construction therefore, 16 more tenants were needed. A time delivery of three months per 8000 m^2 of floor area was scheduled for customisation (i.e. tenants' fitting-out works) in order to meet particular business needs.

At the beginning of 1997, the real estate owner concluded that the tenant market for commercial buildings in Stockholm city was starting to look economically promising. Vacancies in the connected office building were approximately 2%. Revenues were rising and this was regarded as a promising trend. Market analysis was performed continuously as a standard activity, but these were not incorporated in decision-making notes.

Marketing activities conducted during the autumn of 1997 included the fitting-out of showrooms, provision of a website on the internet, advertising in newspapers, and direct contacts with prospective tenants. In the spring of 1998, telephone marketing and visits to possible customers supported these activities. In that year the marketing prospectus focused on (1) IT facilities, (2) high standards of building performance, and (3) tenant freedom in the detailed planning and design of offices. By June 1998, 28 firms were regarded as prospective tenants, but no agreement was reached with any of them.

Eventually the first rental agreement was concluded, in December 1998. The importance of this tenant in bringing life to the entrance area was recognised, as the tenant had many business visitors. This would have a positive effect in giving the large entrance area a lively profile. At this time, an internal customer (i.e. a division of the real estate owner) also entered into a contract. This was regarded as extremely important for marketing purposes. This second contract was negotiated in March 1999, followed by 10 contracts during the remainder of the year; the rental of the last office was agreed in June 2000. The total number of tenants was 14 (15 if the internal contract is included).

The tenants' objectives for their choice of offices can be described as follows.

- Freedom in planning and designing offices.
- Good value for money.
- Political choice for business exposure.
- Fast and timely delivery of office space.
- Perceived collaborative behaviour of the real estate owner (client).

In addition to this list of requirements, tenants said that their working environment was perceived as essential to business as well as offering a competitive advantage when hiring new employees. Tenants also felt that the environment should support their business operations as well as employee performance at work. Furthermore, tenants felt that friendly and collaborative behaviour on the part of the real estate owner supported the process of selecting their future office location.

Communication within the project

One objective of the construction process at Oxenstiernan was to achieve an improved flow of information. This was regarded as a continuous process from planning to on-site works and then on to an administrative service for dealing with future needs in upgrading and the redesign of offices. An integrated project database for all project participants was set up. The overall aim was to reduce lead-time from printing activities (i.e. minutes from different meetings) and the transfer of production drawings between the design office and the site.

After six months of on-site work it was noticed that there was a need for more integrated collaboration in the project. The contractor had to adapt his methods, so that the speed of information transfer, as well as flows between the salesperson, construction project manager and client project manager, could be increased. These three disciplines formed an integrated, working team six months after on-site work began.

The effect of client/tenant objectives and operations in construction

Construction and marketing activities were prolonged for one year beyond the original plan of 18 months. The project prolongation was not considered a problem in the client/contractor relationship. Instead, conflict arose over the process itself.

The contract stated that two offices could be refurbished simultaneously. Nevertheless, because of the locations of the first two tenants, refurbishment work had to be conducted on the top floor and at the entrance level. There was no written agreement on how tenants' fitting-out work should be followed. According to oral statements by the two original project managers (the client's and the contractor's) a top-down approach was assumed to be reasonable for operational purposes.

The impact on the production operations was twofold. First, the construction elevator, intended to support materials transportation for the building, had to be removed ahead of time. Second, materials that had to be changed or added (in response to tenants' fitting-out requirements) had to be carried physically by operatives. This affected the logistics of production, as there were continual changes to the offices' design. Progress of the work overall was delayed because the contractor felt that the logistics and materials handling arrangements were inefficient. Also, from the beginning of the project, the contractor felt that clarifications to shop drawings were late in arriving. Additionally, the contractor reported a number of unplanned stops and starts during construction, affecting the whole office building. For instance, plans allowed for a certain number of toilets in the building and type of floor covering to the offices. A change in the number of toilets and floor covering necessitated additional work and/or re-work. It was felt that decisions on these changes were late and difficult to obtain.

Toilets and office flooring were specified in the main part of the contract, meaning that these works should be completed before fitting-out works began. Later on, tenants' requirements were the subject of change orders. These types of changes were recognised as having a negative effect on the logistical and material flows for the works as a whole.

The contractor claimed that the ideal approach to the works would be top-down: first, the structural frame and core of the building, then services installations. After completion of these elements, fitting-out of offices would similarly adopt a top-down approach. The contractor considered that such a strategy would benefit from repetitive work. This approach was considered to represent effective construction management. Consequently, the contractor regarded the marketing approach as an inefficient project.

Findings

Real estate management tends, to a large extent, to concentrate on physical building conditions and functions in determining project performance. One reason for this could be financial interest in making projects more profitable. In this case study, business actions in refurbishment involve strategic and political objectives as well as those of a more operational nature (Mahmoud-Jouni 2000).

Often missing from portrayals of the construction process is the iterative nature of pre-design and redesign in response to customer needs. Moreover, planning and production processes, in terms of the link between contextual uncertainties and their effect on operational uncertainties, are seldom reported for refurbishment projects.

The Oxenstiernan project started in a traditional fashion. Along the way the product style required for the offices changed. The client decided to upgrade the offices from *normal* standards to what was perceived to be the product requirement among tenants. This change might partly be explained as project uncertainty due to information imperfections in tenants' preferences. This affected project performance, that in turn had an impact on production operations. The empirical findings strongly support the idea of contextual uncertainty and its relation to uncertainties in operational management in construction (Christensen & Kreiner 1991).

It has also been seen how the project environment (i.e. the tenant) affected the works. This might imply a poor briefing process. However, it is also likely that changes in plans might be the consequence of *learning by doing* in refurbishment practice, suggesting that the information content of the project (described as marketing proficiency) and its effects upon production operations management for building works *increase* with project time. This means that project strategy emerges as the response to a specific project environment. By increasing project flexibility the real estate owner chose to reduce project contextual uncertainty. This behaviour also reduced operational predictability and as a result increased operational uncertainty in production operations. At a time when operations management expected to reduce operational uncertainty, it had to handle a situation of increased uncertainty. The result of these interdependent activities became visible as unexpected strategies.

Finally, with the broader perspective described in this case, it could be seen that *contextual uncertainty* and *operational uncertainty* exist both from the viewpoint of the client and the contractor. For instance, from the client's viewpoint there is (contextual) uncertainty in the market for offices, including competition for office space. Operational issues, such as finding a competent contractor and the timely delivery of completed offices, are matters of *operational uncertainty* from the client's perspective. The *operational uncertainty* of the contractor is typically recognised as production operations. Examples of *contextual uncertainty* from the contractor's viewpoint can be recognised as project changes and unexpected technical problems.

Discussion and conclusions

This chapter has examined the conditions of project work in the context of the refurbishment of a building. A study has been carried out from the viewpoint of the owner. The theoretical framework consists of theories of new product development, theories of uncertainty and theories of product assessment. The primary obstacles to greater efficiency in the sector are uncertainty and the institutionalised, fragmented and sequential construction process. Product assessment of project outcomes is considered narrowly in terms of project costs and project time.

This chapter highlights three emergent factors addressing shifts within the real estate and the construction sector. First, there is a *shift in the construction process*. The traditional perspective of the process assumes that information in a project flows linearly. The idea is that you plan for certain conditions (i.e. what planners refer to as *what can be expected*). The idea is then transferred to standard operating procedures in production terms. This is reflective of a passive work pattern. An active pattern would see continuous actions being taken to develop the full potential of a product. This pattern requires everyone in planning and construction operations to interpret environmental changes to adapt the variables of the new product. The passive model, that has been long used in planning and production processes, may benefit from the new product development stream which provides process alternatives, based on active participants and balance between the working organisation and the usability of different production styles.

The second factor involves a *shift to identifying competitive advantage*. Keeping a construction project on the right business track means matching environment changes. Addressing the relationship between process tasks in production and product targeting involves strategies of:

- identifying customers and offerings;
- identifying product advantages; and
- adapting and developing construction process strategy.

The visible attributes that demand a shift in the planning and production processes are identified as increased construction costs, decreasing productivity ratio and client/contractor conflicts. Other attributes can be seen as results of failures in product targeting and of communication failure. This leads to an emergent awareness of the third factor, which is *project outcome and organisation perspective*. Different viewpoints in a project may appear paradoxical in terms of product assessment. Project outcomes can be considered as effective (i.e. the number of tenant contracts) and, at the same time, inefficient (i.e. operations on-site). This suggests that there is a need to analyse the conditions from an organisational perspective as a response to project uncertainty. This may demonstrate that process conflicts are an interdependent activity between the client and the building contractor.

References

Acona, D. & Caldwell, D. (1992) Bridging the boundary: External activity and performance in product development, *Administrative Science Quarterly*, **1**, *4*, pp. 634–65.

Anheim, F. (2001) *A Learning Contractor* (in Swedish: *Entreprenörens lärande)*, Licentiate Thesis, Luleå: Luleå University of Technology (in Swedish).

Anumba, C. & Evbuomwan, N.F.O. (1997) Concurrent engineering in design-build projects, *Construction Management and Economics*, **15**, pp. 271–81.

Brown, S.L. & Eisenhardt, K.M. (1995) Product development: Past research, present findings, and future directions, *Academy of Management Review*, **20**, pp. 343–78.

Christensen, S. & Kreiner, K. (1991) *Projektledelse i löst koblede systemer* (roughly: Project Management – Management and Learning in an Imperfect World), Lund: Academia Acta.

Duncan, R.D. (1972) Characteristics of organizational environments and perceived environmental uncertainty, *Administrative Science Quarterly*, **17**, pp. 313–27.

Eriksson, E. (1994) *Byggbeställare i brytningstid – Bostadssektorn och statligt byggande under miljonprogramsperioden* (roughly: A purchaser in transition, housing and government construction during the one million housing period), Stockholm: Byggforskningsrådet.

Evbuomwan, N.F.O. & Anumba, C.J. (1996) Towards a concurrent engineering model for design-and-build projects, *The Structural Engineer*, **74**, *5*, pp. 73–75.

Fenn, P., Lowe, D. & Speck, C. (1997) Conflicts and dispute in construction, *Construction Management and Economics*, **15**, pp. 513–18.

Galbraith, J. (1973, 1993) *Competing with flexible lateral organizations* (revised edition of *Designing complex organizations*, 1973), Reading, MA: Addison-Wesley.

Holm, M.G. (2000) Service management in housing refurbishment: a theoretical approach, *Construction Management and Economics*, **18**, pp. 525–33.

Jauch, L.R. & Kraft, K.L. (1986) Strategic management of uncertainty, *Academy of Management Review*, **11**, pp. 777–90.

Kadefors, A. (1992) *Kvalitetsstyrning och kommunikation i byggprojekt – analys av ett praktikfall* (roughly: Quality management and communication in construction projects – a case analysis), Report 34, Gothenburg: Chalmers University of Technology.

Katz, R. (1997) Managing Professional Careers: The Influence of Job Longevity and Group Age, in Tushman, M.L. & Anderson, P. (ed.), *Managing Strategic Innovation and Change: A Collection of Readings*, New York: Oxford University Press, pp. 183–99.

Kreiner, K. (1995) In Search of Relevance: Project Management in Drifting Environments, *Scandinavian Journal of Management*, **11**, *4*, pp. 335–46.

Krishnan, V. & Ulrich, K.T. (2001) Product Development Decisions: A Review of the Literature, *Management Science*, **47**, *1*, pp. 1–21.

Lawrence, P. & Lorsch, J. (1967) *Organization and Environment: Managing Differentiation and Integration*, Boston: Harvard University, Graduate School of Business Administration, Division of Research (cited in Duncan, R.D. (1972)).

Liu, A.M.M. & Walker, A. (1998) Evaluation of project outcomes, *Construction Management and Economics*, **16**, pp. 209–19.

Mahmoud-Jouni, S.B. (2000) Innovative supply-based strategies in the construction industry, *Construction Management and Economics*, **18**, pp. 643–50.

Nadler, D.A. & Tushman, M.L. (1997) A Congruence Model for Organization Problem Solving, in Tushman, M.L. & Anderson, P. (ed.), *Managing Strategic Innovation and Change: A Collection of Readings*, New York: Oxford University Press, pp. 159-71.

Nonaka, I. & Takeuchi, H. (1995) *The Knowledge Creating Company*, New York: Oxford University Press.

Pietroforte, R. (1997) Communication and governance in the building process, *Construction Management and Economics*, **15**, pp. 71–82.

Preece, C. & Male, S. (1997) Promotional literature for competitive advantage in UK construction firms, *Construction Management and Economics*, **15**, pp. 59–69.

Robson, C. (1993) *Real World Research: A Resource for Social Scientists and Practitioner-Researchers*, Oxford: Blackwell Publishers.

Sandström, U. (1994) *Mellan politik och forskning* (roughly: Politics and Research), Stockholm: Byggforskningsrådet.

Schuman, P.A., Ransley, L. & Prestwood, C.L. (1995) Measuring R&D Performance, *Industrial Research Institute*, May-June, pp. 45–54.

Song, M. & Montoya-Weiss, M.M. (2001) The effect of perceived technological uncertainty on Japanese new product development, *Academy of Management Journal*, **44**, *1*, pp. 61–80.

Thompson, J.D. (1967) *Organisations in Action*, New York: McGraw-Hill.

Ward, S.C., Curtis, B. & Chapman, C.B. (1991) Objectives and performance in construction projects, *Construction Management and Economics*, **9**, pp. 343–53.

Winch, G., Usmani, A. & Edkins, A. (1998) Towards total project quality: a gap analysis approach, *Construction Management and Economics*, **16**, pp. 193–207.

Womack, J.P., Jones, D.T. & Roos, D. (1990) *The Machine That Changed the World*, New York: Macmillan Publishing Company.

Yin, R.K. (1981) The Case Study Crisis: Some Answers, *Administrative Science Quarterly*, **26**, pp. 58–65.

Yin, R.K. (1993) *Applications of case study research*, Applied Social Research Methods Series, **34**, London: Sage Publications.

Chapter 17

Improving Project Efficiency through Process Transparency in Management Information Systems

Christian Lindfors

Introduction

The construction sector is experiencing large and radical changes forced upon it by external influences, as well as from developments within the sector itself. Examples of external influences are changing social patterns, internationalisation, growing environmental awareness, the rapid development of the IT sector, increasing international competition and more knowledgeable and demanding customers. The need for organisations to adapt to this new set of circumstances is evident as companies face increased competition, while at the same time having to find innovative solutions and increase their customer focus.

One way of adapting, and the one explored in this chapter, is organisational process-orientation. Process-oriented organisations are resource- and time-efficient, and agile to the point that they are able to respond to customer demands and expectations. Instead of having activities aligned according to functions, these organisations are aligned along value chains* of products or product families.

The concept of process-orientation addresses several areas of improvement and is considered to be a useful way of improving efficiency in both projects and organisations. The question of whether or not a process-orientation affects the organisation in a positive manner is still to be explored. The hypothesis is that improved process quality in management information systems will improve project performance with regard to system quality, information quality, process quality, system use, user satisfaction and individual performance. The answer to the question is of strategic importance to the construction sector and its clients in the context of improving overall performance in the sector.

*A firm's value chain is a collection of activities that are performed to design, produce, market, deliver and support its products (Porter 1985).

This chapter highlights the importance of process-orientation within the house-building sector. A state-of-the-art review explores the concept of process-orientation and the different categories of research being conducted in the subject area, including that of the author. The influence of management information systems in achieving project success receives particular attention.

State-of-the-art review

In construction, the concept of value chain management is relatively new: in manufacturing, it is an imperative. Individual actors in the construction process claim to have adopted a value-adding approach to their management of the supply chain, but there are too few to make a meaningful overall improvement. The construction sector lacks this holistic perspective and the co-ordination needed to improve what we might term *the product value chain.*

Process-orientation

Process-orientation of a business can be explained by a change of focus, i.e. adapting a process view of the organisation instead of a functional view. Davenport (1993) says:

> 'a process-orientation to a business involves elements of structure, focus, measurement, ownership, and customers.'

Applying this new perspective to the organisation decreases difficult functional handovers and improves co-ordination capabilities along the product value chain. This implies that an organisation's different workflows and processes are identified and modelled. The working definition in this chapter is:

> 'A process is a collection of activities that takes one or more kinds of input and creates an output that is of value to the customer'. (Hammer & Champy 1993)

A focus on processes, or collections of tasks and activities, that together transform inputs to outputs, allows organisations to view and manage materials, information, and people in a more integrated way (Garvin 1998). Systematically identifying processes within a project organisation and particularly the interaction between them is therefore essential. The central characteristic of a process is that it is a repetitive standardised flow, i.e. it is performed multiple times. Mapping processes makes the dependencies between activities clearer, laying a foundation for organisational development and strategic manage-

Table 17.1 Definitions of approaches performing organisational development with a process focus.

TQM	'A process which ensures maximum effectiveness and efficiency within a business and secures commercial leadership by putting in place processes and systems which will promote excellence, prevent errors and ensure that every aspect of business is aligned to customer needs and advancement of business goals without duplication or waste of effort.' (Pike & Barnes 1993)
BPR	'A fundamental rethinking and radical design of business processes to achieve improvements in critical, contemporary measures of performance, such as cost, quality, service, and speed.' (Hammer & Champy 1993)
Supply chain management	'An integrated philosophy to manage the total flow in a supply chain from supplier to end customer.' (Paulson *et al.* 2000)
Learning organisation	'An organisation skilled at creating, acquiring and transferring knowledge, and at modifying its behaviour to reflect new knowledge and insights.' (Garvin 1993)
Lean production	'Lean production is "lean" because it uses less of everything compared to mass production – half the human effort in the factory, half the manufacturing space, half the investment in tools, half the engineering hours to develop a new product in half the time. Also, it requires keeping far less than half the needed inventory on site, results in many fewer defects, and produces a greater and ever growing variety of products.' (Womack *et al.* 1990)

ment decisions. Process-orientation, then, deals with designing and improving the standardised flow, which also makes it easier to measure (Nilsson 1998). Included in the broad concept of process-orientation are a number of different approaches that adopt a process perspective on organisational development – see Table 17.1.

The ISO 9000: 1994 standard was developed as a system aimed at supporting the achievement of total quality in organisational development. The first version of the standard, however, had the weakness that, if a poor quality product or service was delivered, the standard would see that it was delivered well. In the new and improved 2000 version of ISO 9000, eight quality management principles have been identified to form the basis of the quality management system standards within the ISO 9000 family. These are:

- customer focus;
- leadership;
- involvement and commitment of people;
- process approach;
- system approach to management;
- continual improvement;

- factual approach to decision-making; and
- mutually beneficial supplier relationships.

Top management can use these principles in order to lead the organisation towards improved performance.

The following sections explain the eight principles of quality management in more detail, and use literature from the above mentioned approaches to explain the key issues. One can begin to understand the key issues needed to create a professional process-oriented champion, by using the best examples from the various fields/approaches – see Table 17.2.

Customer focus

Organisations depend more and more on their customers, and therefore should understand present and future customer needs. They should also meet customer requirements and strive to exceed expectations. Processes are the structure by which an organisation does what is necessary to produce value for its customers. Table 17.2 recognises the absolute concordance between the different approaches when it comes to customers. The importance of processes in that they all start with the needs of the customer makes customer satisfaction an important measure to follow. By analysing the process from a customer perspective, the internal ability to increase customer value will be enhanced. Setting the goal of fulfilling customer demands and needs for the lowest price through continual improvement in which everyone is involved will go some way towards satisfying customers (Bergman & Klefsjö 1995).

Leadership

Leaders create a common vision that can unify an organisation by establishing unity of purpose and direction. Without direction, an organisation can pull itself apart. The root cause of many problems plaguing organisations is the missing commitment of their leaders (Drucker 1988). Leadership is becoming more important at a time when information is overflowing in the workplace and transparency is lacking. A crucial task for leaders is, therefore, to form and motivate the internal environment in which people can become fully involved and motivated towards achieving the common goals of the organisation – to act more as a coach than a figure of authority. This makes it essential to form clear strategies and processes, and be able to understand how to transform visions into goals. By drawing attention to processes, the focus is transferred from the finished result to the activities forming them. Since processes create the result, they are also the first things that have to be controlled and developed.

Table 17.2 A matrix mapping the different approaches of process-orientation according to the eight principles of quality management from the ISO 9000:2000 standard.

Approaches / Principles	**TQM and ISO 9000:2000** (Bergman & Klefsjö 1995; Deming 1986; ISO 9001: 2000; Rummler & Brache 1995; Pike & Barnes 1993)	**BPR and process innovation** (Davenport 1993; Hammer & Champy 1993; Hammer & Stanton 1999)	**Value chain management and supply chain management** (Fisher 1997; Paulson *et al.* 2000; Porter 1985)	**Knowledge management and learning organisation** (Garvin 1993; Argyris 1991)	**Lean production and lean enterprises** (Womack *et al.* 1990)
Customer focus	Customer needs, customer requirements, customer expectations	Customer satisfaction, customer value	Customer values	Customer co-operation	Customer relations
Leadership	Purpose and direction	Vision, purpose, mission, process ownership	Goal, strategy	Vision, motivation	Direction
Involvement and commitment of people	All levels, full involvement	Process teams, steering committees	Temporary taskforces	All levels, teamwork	Directed teamwork
Process approach	Activities and related resources are managed as a process	Organisation is managed as processes	Focuses on flows of activities	Well understood process structure	Value streams for products are identified and co-ordinated
System approach to management	Managing interrelated processes as a system	Process teams replace the departmental institutions	Holistic view on activities, chain integration	Management structure that works with teamwork	Managing interrelated processes as a system
Continual improvement	Permanent objective	One time, redesign or innovation instead of improvement	Ad-hoc procedure	Commitment to learning instead of continual improvement	Pursue perfection Kaizen
Factual approach to decision making	Analysis of data and information	Delegated, decision support tools	Analysing cost and quality	Delegated analysis of data and information	Delegated analysis of data and information
Mutually beneficial supplier relationships	Enhances the ability of both to create value	Co-operation in process redesigning	Co-ordination with suppliers	Productive relationships with clients and customers	Collaborative, selected supplier relations

Involvement and commitment of people

Without people all organisations are little more than a name and some fixed assets. People are the heart and soul of an organisation and their full involvement, at all levels, enables organisations to benefit from the fruits of their commitment. Gaining the commitment of the people who work in the process and the managers who oversee them is paramount. Thus, people at all levels in the organisation must be stimulated to master the ability to perform effective teamwork, form valuable relationships with suppliers and customers, critically reflect on their own organisational practices and then change as required. Change in this context means reflecting critically on their own behaviour, identifying their work processes and then changing how they act.

Process approach

A process approach places a relatively strong emphasis on improving how work is done, in contrast to a product- or service-focused delivery, which places most of its energy on improving the product or the service. Adopting a process view of the organisation is a key aspect of process-orientation. In this respect, a process approach starts with the identification of processes, i.e. mapping activities and their interrelated dependencies. A process map can be seen as a blueprint for action and is the foundation for development and improvement-related work, leading to new or improved working methods. The results are often shorter cycle times, better quality, faster results and fewer defects. This structural element of processes is key to achieving the benefits of process-orientation. By making the organisation transparent, deviations and problems are exposed. From these results, processes can be changed, improved and redesigned to eliminate causes leading to deviations.

System approach to management

Managing interrelated processes as a system creates a holistic view and contributes to the organisation's effectiveness and efficiency in achieving its goals. With this new perspective, managers can optimise the organisation from a holistic perspective and focus on broad, inclusive processes. Identifying the activities forming the organisation, and the resources needed, makes it easier to improve the overall performance of the organisation. This leaves room for reflection and analysis, time to think about strategic planning, dissection of customer needs, assessment of current work systems, and inventing new products. Another powerful level is to open up boundaries and stimulate the exchange of ideas. Boundaries inhibit the flow of information: they keep individuals and groups isolated and reinforce preconceptions. Work integration and co-ordination is highly pertinent to organisational success when trying to achieve a common goal and an interpretable strategy.

Continual improvement

A process-orientation encourages organisations to strive continuously to fulfil customer demands and needs for the lowest price through initiatives aimed at improvement in which everyone is involved (Bergman & Klefsjö 1995). There are many different process approaches, each involving a different procedure for improving the process (see Table 17.3). Continual improvement through regular self-assessment, real quality and customer satisfaction can be achieved. An approach to handling regular improvements has to be anchored within the organisation and should include structure, routines, tools and methods. Continual improvement and a commitment to learning should be firmly defined goals within any organisation. The implied identification and close monitoring of these processes would enable the organisation to take corrective and preventive action to eliminate discord and prevent recurrence. Corrective actions take the form of analyses and evaluations of process discordance following process measurement, organisational audits and customer complaints. Preventive actions take the form of extensive analysis of external and internal data and information to prevent surprises and eliminate risk. Most of these kinds of improvement can be handled in prioritised R&D projects to guarantee financial support for corrective measures.

Factual decision-making

Transparent and clearly structured processes are open to measurement in a variety of ways. Such processes can be measured in terms of the time and cost required for their execution. Analysis of data and information should form the basis for all factual decisions and include data collected as a result of monitoring and measurement from all relevant sources. Decisions should be made as close to the source as possible to decrease response times; for example, delegating decision-making responsibilities from the increased use and sophistication of decision support tools. The organisation shall decide, collect and analyse appropriate data to demonstrate their suitability and effectiveness and to evaluate where continual improvement is best suited. The analysis of data should provide information relating to customer satisfaction, conformity to product requirement, characteristics and trends in processes and product, including opportunities for preventive action, and suppliers.

Table 17.3 Different approaches to process improvement.

Garvin 1993	Systematic problem solving	Experiment with new approaches	Learning from experience and past history	Learning from experience and past history	Transferring knowledge throughout the organisation
Hammer & Champy 1993	Scope project	Learn from others	Create *To-Be* processes	Plan transition	Implement new process
Davenport 1993	Identify processes for innovation	Identify change levers	Develop process visions	Understand existing processes	Design and prototype new processes
Deming 1986	Study the process	Change the process on a small scale	Observe the effects	Analyse results	Study the process
ISO 9001 2000	Establish the objectives	Establish the processes	Implement the processes	Monitor and measure processes	Continually improve process performance
Rummler & Brache 1995	Project definition	Process analysis and design	Implementation of processes	Evaluate process performance	Identify and implement improvements

Mutually beneficial supplier relationships

An organisation and its suppliers are mutually dependent and a mutually beneficial relationship enhances the ability to improve products and processes, as well as create value for both parties. Collaborative approaches are therefore a natural way of handling supplier relationships. Doing so increases the organisation's ability to achieve a culture of continual improvement and the ability to co-ordinate inter-organisational processes, as well as raising process and product quality.

Management information systems

Process-orientation initiatives have tended to be supported by paper-based process models and routines, describing the dos and don'ts. Progressive computerisation of processes has increased the flow of information, but not necessarily improved efficiency and cost-effectiveness. Even so, computerisation is an essential component for improving flow efficiency in a chain (Paulson *et al.* 2000). By using management information systems as an implementation vehicle, organisations can integrate process work into routine daily work naturally and cope with more information. Moreover, the internet can be used as an interface for the development of management information systems across the whole value chain. Understanding and identifying those factors that contribute to management information system success is therefore a necessity when speaking of evaluating such an initiative.

DeLone & McLean (1992) introduce a descriptive taxonomy of dependent variables to measure information system success. The information system success model – see Fig. 17.1 – consists of six interdependent constructs (information quality, system quality, use, user satisfaction, individual impact and organisational impact), that is an attempt to reflect the process nature of information system success.

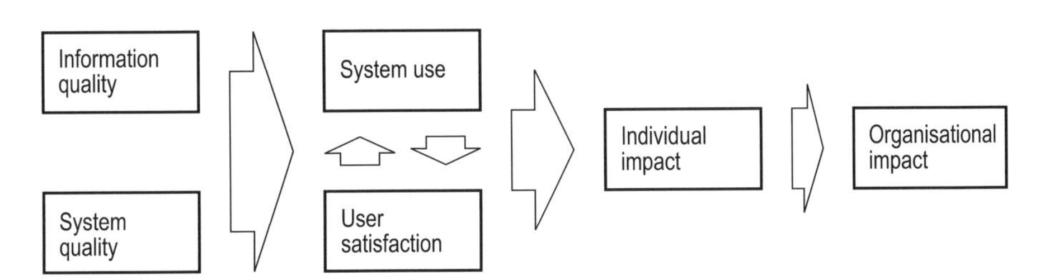

Fig. 17.1 The six main factors of the information system success model (DeLone & McLean 1992).

Process-orientation vs. management information system success

Two different factors share a major role in improving project performance. Information is not data in a process-orientation, but a structured way of describing informational needs of managers. Process initiatives have a significant impact on improving project understanding, but are merely descriptions of projects. Identifying processes and then instantiating them in everyday work provides therefore one means of improving work efficiency. Information technology is considered to be an excellent enabler, although not a driving force for implementation.

A value chain contains both the physical flows and the information flows. By using this definition, one can infer that the information flow affects the physical flow in a direct way. As an example, an incomplete, sporadic and tardy information flow creates the need for large safety storage in the different phases of the chain, while a complete, frequent and current information flow creates opportunities for lowering storage levels (i.e. inventory). Thus, the main function of the information flow is to administer the physical flow and make it efficient (Paulson *et al.* 2000).

APQC (American Productivity and Quality Center) suggests that an integrated, holistic and systematic approach will be essential to organisations in the future. Quality will be used as it was initially intended – as an integral part of the business management philosophy and organisational fabric (Cates *et al.* 2000). According to Gilchrist & Kibby (2000), major enterprises are looking for greater business effectiveness through improved information access; or efficiencies in information retrieval, which free time for more productive activities. One of the answers to information overload might be a corporate taxonomy providing a knowledge map for information and communication channels and a roadmap of the intellectual capital of the company.

Creating a stable corporate intranet with a dynamic interface to a project management information system would enable the user to gain an understanding of available information and the means for accessing it. To develop a system that integrates the whole chain, and that uses a process perspective, would imply a completely new way of thinking and pose a number of questions concerning implementation. By combining the *information success model* with factors of process quality (measures of the information management process), expectations of unravelling the impact of process-orientation on management information system success would seem rightly optimistic.

Research project

Project description and objectives

The aim of the project is to evaluate today's house-building process from an information management perspective and, together with the sector, produce and establish a coherent process model.* Based on this, advantages gained from the introduction of a transparent business process coupled with a newly developed project management system (or management information system) will be assessed. The result will lead to a greater understanding of how to improve project performance by successful management information systems, and how to manage projects better, cheaper and faster in the future.

Research methodology

The research aim is to evaluate the impact of process quality on the success of management information systems. The research process follows an exploratory approach. A process model was derived from empirical data captured in interviews, a case study and questionnaire surveys. After evaluating process-modelling alternatives, a model was produced using the IDEF0 method (Karhu 2000; Malmström *et al.* 1998). Based on the process model, a process-oriented management information system has been developed that represents activities, workflows, documentation, information, communication and participation.

To understand the factors that are typical for process quality, an extensive literature study was undertaken. Among the different approaches, a matrix was created, listing the main variables of process quality. As a result of the literature study, a framework was created to test the hypothesis:

> 'improved process quality in management information systems will improve project performance with regard to system quality, information quality, process quality, system use, user satisfaction and individual performance.'

The framework draws upon the model used by DeLone & McLean (1992) in evaluating information system success (see Fig. 17.2), adding six variables of process quality to test the influence of process-orientation on project performance with management information systems.

*NCC Housing's Total Package Concept stands as a model of how it is performed today. The Total Package Concept is NCC Housing's name for development/construction promises covering the process from the project idea to customer support.

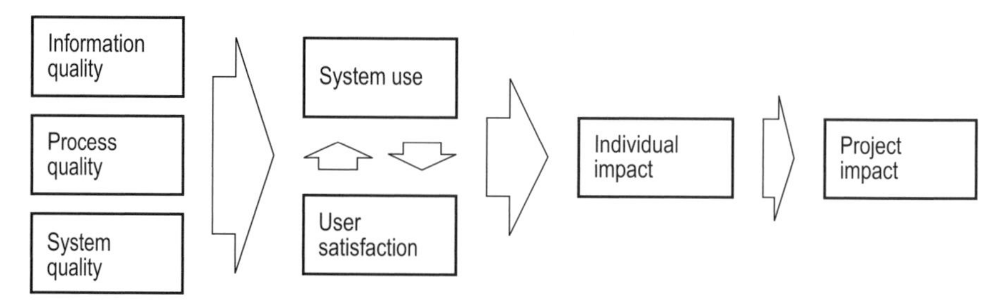

Fig. 17.2 The research hypothesis to be tested (augmented from DeLone & McLean's (1992) information system success model).

A questionnaire-based survey has been adopted to test the complex relationship between the different underlying measures of the hypothesis. The survey is intended to measure attitudes of project managers (i.e. for the purpose of homogeneity) using the newly developed management information system. For studies of attitude patterns or explorative theories of attitudes, the Likert procedure is considered relevant (Oppenheim 1966).

The following statement is an example drawn from the questionnaire survey: *I find the access capabilities of the management information system to be convenient.* Responses are graded with a seven-item Likert-scale, rating the responses from *strongly agree* to *strongly disagree.* All questions are very direct, making them more uniform and the results more reliable. The results of the questionnaire survey will be statistically analysed and an improved model will be presented. The methodology adopted for the research has been exploratory at the early stages and then has used attitudinal research to gain deeper understanding. The method adopted has been, therefore, the generation of theory by interpretative case studies, surveys and interviews within a construction organisation, and then the use of a comprehensive questionnaire to find answers to the underlying research questions.

Conclusions

This chapter's scope shows that the use of an accurate representation of the project process, with the help of a management information system, will help to improve its performance significantly. A process-orientation approach is presented as a solution for improving project efficiency. Using the six variables of process quality, it is possible to test the principal hypothesis in the research. Measuring the attitudes of a homogeneous group increases confidence in the ability to improve understanding of management information system success factors.

One of the main prerequisites for any substantial improvement within an organisation is to understand the *bigger picture*. Focusing on the details might well cause problems somewhere else in the organisation. The use of management information systems as an implementation vehicle for process-orientation initiatives is also crucial when trying to handle enormous amounts of information. However, without the tool it would have been almost impossible to structure all the information.

Thoughts about future research could be expressed in the simple question: *how do we organise to produce the best possible results in the future and what measures do we have to take to realise them?* Analysing the results from the intended questionnaire survey provides a good foundation for future study. The next step in the research will be to use the result and extend it to the whole organisation.

References

Argyris, C. (1991) Teaching smart people how to learn, *Harvard Business Review*, May–June, Harvard Business School Press, Boston, MA, pp. 99–109.

Bergman, B. & Klefsjö, B. (1995) *Kvalitet – från behov till användning*, 2nd edition, Lund: Lund Institute of Technology (in Swedish).

Cates, C., Demery, A. & Hunzeker, R. (2000) *Quality approaches for the new millennium*, Consortium benchmarking study, Best practice report, Houston, TX: American Productivity & Quality Center.

Davenport, T.H. (1993) *Process Innovation: Reengineering Work Through Information Technology*, Harvard Business School Press, Boston, MA.

DeLone, W.H. & McLean, E.R. (1992) Information Systems Success: the quest for the dependent variable, *Information Systems Research*, **3**, *1*, pp. 60-95.

Deming, W.E. (1986) *Out of the crisis*, Cambridge, MA: First MIT Press.

Drucker, P.F. (1988) The coming of the new organisation, *Harvard Business Review*, Jan-Feb, Harvard Business School Press, Boston, MA, pp. 45-53.

Fisher, M. (1997) What is the right supply chain for your product, *Harvard Business Review*, March–April, Harvard Business School Press, Boston, MA, pp. 105-16.

Garvin, D.A. (1993) Building a learning organization, *Harvard Business Review*, July-August, Harvard Business School Press, Boston, MA, pp. 78-90.

Garvin, D.A. (1998) The process of organization and management, *Sloan Management Review*, Summer 1998, **39**, 4, pp. 33-50.

Gilchrist, A. & Kibby, P. (2000) *Taxonomies for Business: Access and Connectivity in a Wired World*, London: Thomson Financial Publishers Ltd.

Hammer, M. & Champy, J. (1993) *Reengineering the Corporation – a Manifesto for Business Revolution*, London: Nicholas Brealey Publishing.

Hammer, M. & Stanton, S. (1999) How process enterprises really work, *Harvard Business Review*, November-December, Cambridge, MA, pp. 108-18.

ISO 9001 (2000) *Quality Management Systems Requirements*, Brussels: CEN (European Committee for Standardization).

Karhu, V. (2000) *Formal Languages for Construction Process Modelling*, Proceedings of International Conference on Construction Information Technology CIT 2000, Reykjavik, Iceland: CIB-W78, IABSE, EG-SEA-A1, pp. 525-34.

Malmström, J., Pikosz, P. & Malmqvist, J. (1998) *The Complementary Roles of IDEF0 and DSM for the Modelling of Information Management Processes*, Proceedings of the Fifth ISPE International Conference on Concurrent Engineering: Research and Applications, Tokyo, Japan, pp. 261-70.

Nilsson, G. (1998) *Process-Orientation, Integration of Work Teams and Management Control*, SSE/EFI Working Paper Series in Business Administration No. 4, Stockholm: Stockholm School of Economics.

Oppenheim, A.N. (1966) *Questionnaire Design and Attitude Measurement*, New York: Basic Books.

Paulson, U., Nilsson, C.-H. & Tryggestad, K. (eds) (2000) *Flödesekonomi – supply chain management*, Lund: Lund Institute of Technology (in Swedish).

Pike, J. & Barnes, R. (1993) *TQM in Action: A Practical Approach to Continuous Performance Improvement*, London: Chapman & Hall.

Porter, M.E. (1985) *Competitive Advantage: Creating and Sustaining Superior Performance*, New York: The Free Press.

Rummler, G.A. & Brache, A.P. (1995) *Improving Performance: How to Manage the White Space on the Organizational Chart*, San Fransisco: Jossey-Bass.

Womack, J.P., Jones, D.T. & Roos, D. (1990) *The Machine that Changed the World*, New York: Rawson Associates.

Chapter 18
Improvement Processes in Construction Companies

Peter Samuelsson

Introduction

Construction has one of the worst public images among the industrial sectors (Santos *et al.* 2000). Nevertheless, this poor image should be an effective motivator for improvement. Competition between construction companies should be another motivator, considering increased internationalisation. A third motivator for a focus on improvement in construction companies is the new ISO 9000:2000 standard, that requires greater effort in bringing about continual improvement than its predecessor from 1994. Hence, to create and uphold improvement processes in a business is crucial; primarily to survive as a successful company in a growing and changing environment, but also to create an image of a best practice company with certificates as proof. An approach to management that exhorts every employee continually to improve everything that he or she does in the interest of customers is required (Tam *et al.* 2000).

Companies generally within the construction sector do not have a reputation for being oriented towards TQM (total quality management) values such as continual improvement. Some still struggle with the concepts, avoiding the use of available tools and techniques, and in so doing deprive themselves of the opportunity to improve their performance (Gieskes & Broeke 2000). Furthermore, projects are usually temporary alliances of autonomous partners, that may imply that participants are partners and in a state of constant competition at the same time (Gieskes & Broeke 2000). Short-term benefits are almost minimal and the strategic component in decision-making is often absent (Gieskes & Broeke 2000; Love *et al.* 2000).

Sommerville (1994) suggests five broad subheadings for the resistance forces specific to the construction sector's adoption of TQM:

(1) product diversity;
(2) organisational stability;
(3) holonic networks and change;

(4) contractual relationships; and
(5) teamwork and management behaviour.

Establishing processes for continual improvement in construction organisations is apparently not easy, yet authors report that it is possible (Sommerville & Robertson 2000; Gieskes & Broeke 2000; Orwig & Brennan 2000). Love *et al.* (2000) suggest that construction organisations must re-think their approaches to TQM in order to learn and change simultaneously. In this context continual improvement (CI) is vital.

The purpose of this chapter is to identify key aspects of the improvement process within construction companies, with respect to the concept of continual improvement. The concept of CI is outlined together with elements in the field of its deployment. Using a state-of-the-art review of current knowledge as the background, real life improvement processes in the construction company Skanska Sweden, a business unit within the Skanska Group, are analysed. It is argued that CI involves organisation-specific behavioural routines that are built up over time, and formalised operating procedures. The analysis extends beyond the formalised procedures in the case study company to include consideration of how procedures are deployed and how behavioural patterns generate improvements.

State-of-the-art review

The TQM philosophy

Improvement of performance is central in the philosophy of TQM, but also in other schools of management thinking such as JIT and BPR (Bond 1999). TQM is considered to be a management paradigm capable of facilitating the attainment of organisation-wide quality awareness and continual improvement (Ghobadian & Gallear 1997). Since the 1980s, TQM has developed into an established philosophy for improving business performance at all levels (McAdam 2000; Hellsten & Klefsjö 2000). Laszlo (1999) concludes that

> 'TQM is the thirst for improvement that goes beyond the focus on doing things right and embraces looking for ways of doing things better'.

With respect to the above, TQM is considered to be an appropriate philosophy to start from when studying current knowledge relevant to the improvement process of a construction company. Hellsten & Klefsjö (2000) define TQM as a management system consisting of three independent, but supportive, components: core values, techniques and tools (Fig. 18.1).

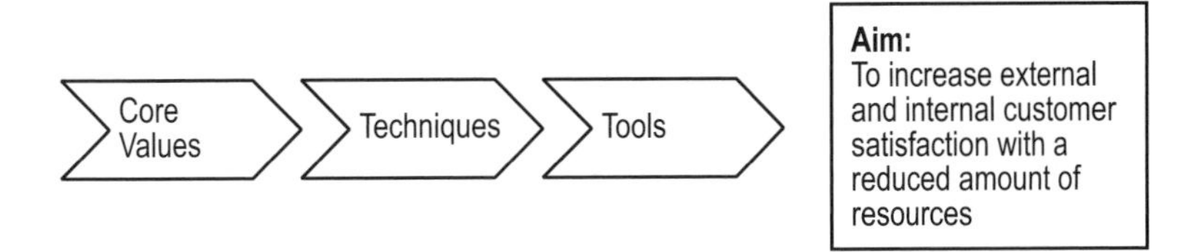

Fig. 18.1 Role of core values, techniques and tools (adapted from Hellsten & Klefsjö 2000).

This simplifies the understanding of the concept and implies that the aim of TQM is to establish a culture based on the core values also defined by Hellsten and Klefsjö (2000), which are:

- focus on customers;
- management commitment;
- focus on processes;
- continual improvement;
- fact-based decisions; and
- everybody's commitment.

In this chapter the focus is on the fourth of these values, continual improvement (CI).

Concepts of improvement

As an aim of TQM is to establish a culture based on CI, it is a mistake to halt efforts for improvement arguing that a process for bringing about improvement has been developed. Hence, those people who refer to TQM as a programme do not understand the fundamentals of CI (Dale & Bunney 1999). To describe CI in short, the definition stated by Savolainen (1999) is used:

> '[CI is] a company-wide process of focused and continuous incremental innovation or improvement'.

Other descriptions of CI do not emphasise the word *continuous* or *continual* as much; Michela *et al.* (1996) describe CI as:

> 'a collection of activities that constitute a process intended to achieve improvement'.

In contrast, a more extensive description by Nicholson *et al.* (1995) is that:

> 'The idea of CI suggests that there may be no state of dynamic equilibrium, rather that organisational members consciously choose to keep the organisation in a chronically unfrozen state'.

Thus, it is suggested that CI is not only a continual, but also a company-wide process, driven by organisational members and, in a sense, a self-generating process.

Several authors list criteria claimed to be crucial for CI – see for example, Harrison (2000), Sun (2000) and Kaye & Anderson (1999). As pointed out by Kaye & Anderson (1999), it takes more than a static list of criteria to find the key to continual and self-generating improvement. Michela *et al.* (1996) even claim that it is impractical to provide lists of activities entailed by CI or conditions for its success, since a shift to CI as a way of working has implications for so many aspects of the organisation (for example, strategy, operations, human resource policies and practices). Instead, they suggest that a key challenge when implementing CI is to ensure alignment or congruence among strategy, operations, human resources and other domains of management practice. A better understanding of this synergy is argued to be helpful for improving practice in CI (Michela *et al.* 1996).

Harrison (2000) claims that CI must be viewed in the context of other features of a socio-technical system. According to his research, CI emerges at the operator level as a fragile *human-ware* category whose position relative to other categories is strongly influenced by the type of operating system in use, and its accompanying knowledge base. Bessant & Francis (1999) seem to have a similar conception, suggesting that CI can be considered an example of dynamic capability. In this context, strategic advantage is seen to come not from simple possessions of assets or from a particular product/market position but stem from a collection of attributes that are built up over time in a highly company-specific fashion. These attributes provide the basis for achieving and maintaining competitive edge in an uncertain and rapidly changing environment. CI represents an important element in such dynamic capability since it offers mechanisms whereby a great deal of the organisation can become involved in its innovation and learning processes. Its strategic advantage is essentially as a cluster of behavioural routines. This explains why it offers considerable, and perhaps inimitable, competitive potential; behaviour patterns take time to learn and institutionalise, and are hard to copy or transfer (Bessant & Francis 1999). Hence, the role of behavioural aspects must be considered when discussing CI.

The principle of CI may appear to be in conflict with project management, since projects produce unique results. By definition, continual improvement of a singular effort is impossible. Additionally, as projects are temporary, measurement and reward systems for project managers tend to be based on short-term measurements of schedule, cost and technical performance that undermine the long-term emphasis on CI. These characteristics are highly visible in the construction sector and in the case study company. Still, if project management is considered an ongoing process in an organisation it is obvious that continual improvement of that activity, despite everything, is possible (Orwig & Brennan 2000). Orwig & Brennan conclude that formal project management is quality management and

> 'quality management fundamentals applied to the project-based organisation is good business'.

Improvement in practice

In considering Fig. 18.1, achieving a culture based on core values requires tools and techniques. Skanska's primary tools and techniques for improvement are their measures, their management system and their internal audits. A brief outline of current views on whether or not such elements in general support or inhibit CI activities in an organisation is needed.

Use of monitoring and measurement are key enablers for CI on higher levels (Bessant & Francis 1999). In a perfect world, a performance measurement system provides early warning, indicating what has happened, diagnosing reasons for the current situation and indicating what remedial action should be taken (Bond 1999). This implies that measurement is a control over activities. In contrast, Bessant & Francis (1999) claim that the purpose of measurement in the context of CI is to enable and monitor the rate and direction of improvement, and not as a traditional behaviour measure, with control over activities. Relevant measures need to be identified and used to gauge the extent to which performance has changed. Harrison (2000) supports this view, recommending focus on measuring process improvement, not results:

> 'setting output targets as the principal measure of CI activity leads to failure of that activity'.

Harrington (1995) claims that CI is inhibited by explicit knowledge-based systems, since the primary aim of such systems is to do the job in the same way, in the same time, every time. In contrast, several authors suggest that the presence of an exhaustive management system, including formalised procedures for project management, supports CI activities (Bessant & Francis 1999; Bond 1999; Kaye & Anderson 1999). Bessant & Francis (1999) emphasise the need for formalised operating procedures, and mechanisms for updating them with the results of CI activity, implying that formalised procedures can be improved by using CI activities. Kaye & Anderson (1999) also emphasise the significance of an operating system, aiming at another function. They claim that a robust framework is required in order to assess and measure performance, which is essential in the context of CI.

An assessment identifies where improvement is needed and therefore becomes a foundation for continual improvement (Orwig & Brennan 2000). Auditing is a way of assessment that traditionally has been viewed as a negative process by its recipients (Beecroft 1996), but now several authors report that the role of internal auditors is changing from a traditional checking approach to a more proactive value-added approach. Bou-Raad (2000) suggests that internal auditors are taking up partnership with management and that an emerging number of organisations appear to be realising that internal auditors can render a better service to management through their increased involvement in business practices.

Karapetrovic & Willborn (2000) have made a similar finding claiming that auditing as a source of information has become very useful in modern business, especially with business' increasing complexities and pressures for continual adaptation and improvement. They suggest that a concept of system-based generic auditing, together with a sound generic guideline, could provide for an integrative approach for management system auditing. In addition, Becket & Murray (2000) report that a participative, yet structured, auditing process can provide a vehicle for organisational learning across the company. Another shift reported regarding audits is that they are becoming more generic, spanning over quality, environmental, safety or financial disciplines (Karapetrovic & Willborn 2000; Bou-Raad 2000). Hence, audits are increasingly being used for the primary purpose of continual improvement and learning in a broader sense, and not strictly compliance to stated requirements. A limiting factor might be the difficulty of obtaining competent people who can undertake a broad type of audit (Bou-Raad 2000).

Research project

Project description and objectives

As stated in the introduction, the purpose of this chapter is to identify key aspects of the improvement process in construction companies, with respect to the concept of continual improvement. It has been suggested that improvement efforts emerge from organisation-specific behavioural routines that are built up over time. However, it was also suggested that formalised operating procedures were significant in the context of CI. Therefore, the aim is to analyse the formalised procedures in the case study company and also to capture how these procedures are deployed and how behavioural patterns are generating improvements. In this way, it is possible to map out the improvement process.

Research methodology

The research calls for a deep understanding of the company's way of working and its culture. Such understanding requires an in-depth study of the company's operations, which is why a case study approach has been adopted. Case studies are distinctive in their ability to attend to programme operation and context, but also to capture processes and outcomes in a causal logic model and thereby provide useful intermittent feedback (Yin 1993). As this project is limited to one particular organisation, the case study is single; still, a case study can develop lessons that can be generalised to the major themes in a field (Yin 1993). A single case study gives the opportunity of in-depth studies in an organisation, especially if it is, as in this case, conducted over some time in close co-operation with the case study organisation. The study is participative, as the researcher is actively involved in the development of tools and methods at the company, therefore the project can be placed in the category of action research (Fellows & Liu 1997).

Close co-operation with the development manager at Skanska has helped to establish an overall picture of the organisation. Primarily, the researcher has attended the meetings of the development group, which includes the development manager of Skanska and correspondent representatives from directly subordinated organisational units. Data from these meetings and individual daily work have been complemented by more structured interviews with line managers, as well as people engaged in supportive functions. Interviews with line managers serve to give a production view of the organisation, which is crucial as the researcher mostly works in a development environment, where issues are seen from a different perspective. The study has, at this point, been running for almost a year and it is planned to continue for another six months.

A benefit of the approach is the opportunity to compare similarities and contrast dissimilarities in what people actually do and how they describe what they do. As a consequence of active involvement in development work at the company, the thin line between observations and interviews is very diffuse. Nevertheless, discussions at meetings held as part of the daily routine can actually be considered as a kind of informal and unplanned interview (Kaijser & Öhlander 1999). Since the method generates a substantial amount of information, of which a proportion will not be relevant to the research question, it is crucial to apply some kind of priority. Therefore the approach to observation must shift between open and focused, depending on the situation. In recognising that the surrounding environment may influence a participating researcher, transcripts from interviews conducted in other studies of the organisation have been used as complementary sources of information.

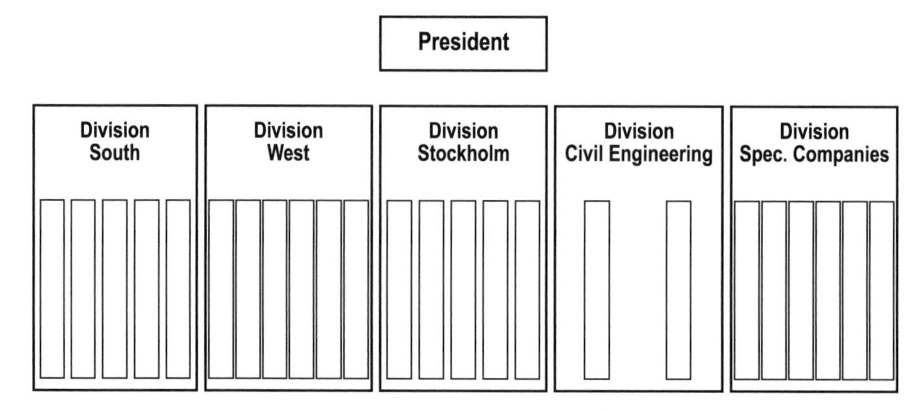

Fig. 18.2 Organisational chart of Skanska Sweden.

Research results and industrial impact

Quantification of results

The case study company has adopted as its mission: to develop, build and service the physical environment for living, working and travelling.

Its business is carried out within Sweden and comprises road construction, building construction, housing and civil engineering projects. Specialist industrial companies are included in a separate division.

The company has approximately 15 000 employees, and it is organised into three divisions covering regional areas of Sweden and two product-specific divisions (Fig. 18.2). The three regional divisions are divided into five or six product-specific regions (vertical bars in figure) consisting of four to eight districts. The number of employees in each district is around 100–200. However, the size of the districts varies considerably and there are some exceptions regarding the generic organisational pattern described here.

The improvement process in the company

The procedures described in the management system of the company imply a structure for the improvement processes as described in Fig. 18.3. The figure represents the improvement process at all units within the company, including districts, even though some of the elements are not executed at each unit. For example, review of the management system is carried out at divisional and company level, but the subordinated units provide input to the review. However, it is apparent this figure does not fully describe the real life situation within the company. Even if practices and tools are generally considered to be

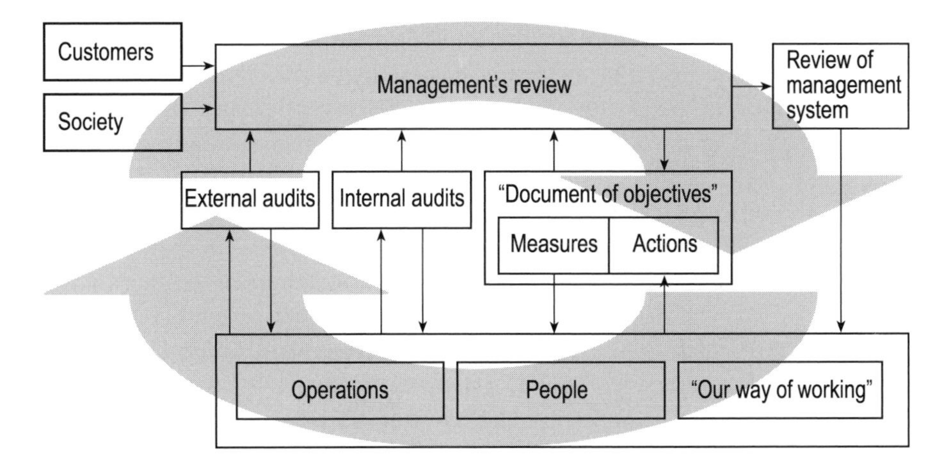

Fig. 18.3 The expected improvement processes.

very useful and well designed by people in the organisation, deployment of them varies among the units and informal efforts influence the processes.

As Fig. 18.3 implies, improvement work is an ongoing activity with no definite beginning or ending. Management's review is conducted on all levels of the organisation and means that information gathered from the organisation and its surrounding environment is reviewed. A significant source of information is the compiled results from internal and external audits; the compilation of the internal audits is, however, conducted in different ways in the divisions, which is why a company-wide exercise is difficult. Other sources of information for senior management's review are measures of several factors and discussions from meetings, but also informal conversations and individual suggestions from co-ordinators and managers.

The most important actions for improvement derived from the information are prioritised and set up in a *Document of objectives*. The *Document of objectives* is a balanced scorecard including objectives, measures and actions in six areas: finance, customer, people, operational development, environment and work environment. Specific *Documents of objectives* are deployed on a yearly basis in all units of the organisation having overall responsibility for results (i.e. districts, regions, divisions and company). The objectives in the six areas are company-wide, but actions for improvement are unit-specific. Based on the *Document of objectives* of the superior organisational level, each unit establishes its own specific actions. This is in accordance with the general view among people in the organisation, which is that planning at higher levels should focus on business policy and strategy as a whole and then provide guidelines on planning to districts. A strong focus on financial and market issues is prominent throughout the organisation, but especially at the district level.

Actions for improvement must be clear and concrete at the district level, but there cannot be too many. Experience shows that actions may otherwise remain in the *Document of objectives* for several years. Actions should be reviewed at management meetings, but the real situation indicates that the review frequency of actions varies considerably among the units. Many of the actions for improvement recur in several of the units, but the exchange of experiences remains rather poor. This may be a result of the entrepreneurial culture of the company – the units are accustomed to managing themselves. A reason for failure to realise actions for improvement expressed by some is that the organisation is too trimmed – people are too busy handling their daily work; therefore, it is hard to drive actions in an effective way. Looking for better ways of accomplishing activities is still not 'a way of life' (Laszlo 1999) for everyone at Skanska. The focus is on 'doing the daily job' and making money, with little effort put into developing and improving the way the job is carried out. However, there are indications that attitudes are changing and senior management's reviews are an attempt at capturing the complete picture of the operations, thereby ensuring alignment or congruence between strategy, operations and other domains of management practice (Michela *et al.* 1996).

Although the *Document of objectives* is the primary tool to plan and steer the company's business, it also has a reporting function, since the measures included provide important data for senior management's reviews. In Fig. 18.3, measures based on *Customers* (i.e. customer satisfaction index) and *Society* (i.e. image index) are viewed separately, while measures based on internal issues within the company are signified by *Measures* under the heading of *Document of objectives*. The measures imply trends and performance in the areas of the scorecard. Most of them are measured annually at company level, but some are division-based. The focus on financial performance is reflected in senior management's reviews of the measures and is reviewed 4–5 times annually at their meetings.

Measurement of other areas is not conducted as frequently. A general review of the measures is, however, carried out on an annual basis. Although review routines vary across the units and the work is perceived as very time-consuming, managers agree strongly that it is important to understand and use the *Document of objectives*. Considering this, it is evident that current practices do not fulfil the requirements of a performance measurement system in the perfect world, as outlined above (Bond 1999), and certainly cannot be seen as measures in the context of CI (Bessant & Francis 1999). Nonetheless, measures for several areas are established and, not least, there seems to be a consciousness of the significance of measurement and review in the organisation. Despite this, measurement is required to a greater extent if processes for continual improvement are to be effective.

The backbone of the organisation's operations is an integrated management system called *Our way of working*. This is a tool for the entire operational sphere and encompasses all projects, irrespective of size and type. Work procedures for the core business of the company (construction projects), as well as leadership and supporting functions such as customer/market, employees, purchasing, technology and product development, are described. Along with these descriptions are aids in the form of templates and checklists. *Our way of working* embodies requirements of the ISO 9000 and 14000 standards and aspects of the internal work environment and offers a broad perspective of the business. Within the organisation, *Our way of working* is seen as a supportive tool for daily work by both management and operatives. A telling description proffered by a member of the organisation is *the handrail on the stairs when the light has gone out*.

Most people are convinced that to deploy *Our way of working* fully on all projects would improve the overall performance of the company significantly. Yet, as *Our way of working* serves as a guide describing the operational procedures, it must be developed further in order to improve operations. It is unfortunate, therefore, that some employees do not feel that they can contribute to development and improvement of the system. Senior management's reviews aim to capture views across the organisation, but suggestions for improvement are not collected in a systematic way. With respect to the above – see earlier section – it is claimed that formalised procedures could be updated from the results of CI activities (Bessant & Francis 1999). This principally calls for two aspects to be considered: first, the procedures of the system must be fully deployed; second, a structured way to assess and improve the procedures described must be established. Possibly, *Our way of working* could also serve as a basis for performance measurement (Kaye & Anderson 1999), but this needs to be examined further.

Conducting *internal* and *external audits* are the main approaches to assessing the operations of the company. There are two types of internal audits in the organisation:

(1) audits of specific construction projects, regarding quality, environment and/or working environment; and
(2) audits on a business unit as a whole.

The explicit purpose of internal audits is to detect areas for improvement and to assess that operations are appropriate, efficient and in accordance with *Our way of working*. Audits are appreciated, as most people think they give a valuable *health check* of their organisation. The content of the audits is considered to be serious and reliable, but it is emphasised that the auditors must be highly qualified in order to make the audit an opportunity for improvement and learning. External audits are required to retain ISO certificates: the results from both external and internal audits serve as basic data for senior management's review.

In the past, many looked upon audits as controlling police activities, but now this attitude has changed, as managers and auditors have declared the purpose of audits more clearly. Consequently, the case study strongly supports the views of authors stated above regarding change in attitude towards audits (Beecroft 1996; Bou-Raad 2000), and increasing extent of audits (Becket & Murray 2000; Bou-Raad 2000; Karapetrovic & Willborn 2000). It is clear that audits are gaining an important role in assessing performance and thereby detecting areas for improvement in Skanska. Nevertheless, the issue of mobilising a sufficient number of qualified auditors (Bou-Raad 2000) is crucial for the company. Even though the internal audits serve as opportunities for improvement at the units, they do not stimulate suggestions for improvement effectively and the units themselves do not drive them. Skanska has therefore evaluated more self-generating approaches to assessment of their operations, such as self-assessment by business excellence models (e.g. Samuelsson & Nilsson 2002). However, these models have been rejected, as they were not considered to be sufficiently aligned with the company's operations.

Implementation and exploitation

The results of this study imply that the management system provides a framework for improvement processes in a construction organisation. Even though construction projects often are unique, elements of them are not. By describing efficient approaches for repetitive elements and prescriptions regarding the external environment and work environment, better conditions for well-functioning operations are created. Deploying *Our way of working* in Skanska is considered to increase the likelihood of success in projects, with regard to financial results as well as customer satisfaction. Assuming that *Our way of working* is fully deployed, a well-functioning process for assessing, reviewing and improving it would imply continual improvement of operations. *Our way of working* is perceived as a useful tool by most people at Skanska, but the deployment of it is not comprehensive and a structured way of collecting suggestions for improvement is missing.

Internal auditing is a way to assess whether procedures are deployed, and is thereby a driver for increased deployment. Disadvantages are that audits require much time from auditors as well as respondents; they are not conducted regularly at all units; and they are not driven by the units themselves, where they can still be seen as controlling activity rather than support for improvement. Audits assess whether recommended approaches are deployed, but it must also be assessed whether those are the most efficient ones. It is suggested that practices to assess deployment of approaches and to collect areas for improvement of the management system should be established. Practice on a voluntary basis may stimulate a culture of continual improvement more effectively than traditional audits.

Conclusions

The construction sector has been shown not to be oriented towards improvement and learning. Production is comprised of projects that are said to be unique and not repetitive and that complicate learning and use of formalised yet simple approaches for assessment, review and improvement. The purpose of this chapter was to identify key aspects of the improvement process in construction companies, with regard to the concept of continual improvement. A participative long-term case study of a large construction company has been conducted in order to identify these aspects.

The case study implies that the sector has a journey ahead to achieve a culture of continual improvement. A key aspect of improvement processes is measurement of performance. In this case several measures are established, but the main focus is on financial issues and other measures need to be conducted and reviewed more frequently, considering the concept of continual improvement. As construction projects are often unique, a key aspect for realising continual improvement is a well-designed management system, describing efficient approaches for repetitive elements of projects. Potential benefits are significant, particularly in large organisations as they have greater resources for maintaining and improving it. However, there must be a structured way of improving the approaches and, of course, the approaches must be deployed. Audits can be seen as a driver for increased deployment and it is evident that the attitude towards them is changing for the better. Still, audits require qualified auditors and they do not stimulate suggestions for improvement from the organisation effectively. Therefore a more self-generating complementary approach for continual improvement is eligible.

To sum up, efforts for a business more oriented towards improvement have been made in construction companies. This study implies that a formal foundation for improvement processes is established, but its full potential is not realised. In order for this to happen, the will to do things better must be stronger in the organisation. As the prevailing main focus is on doing things right, this will not happen by itself. Hard work and more self-generating approaches for improvement are required to achieve significant improvement of performance.

References

Becket, R. & Murray, P. (2000) Learning by auditing – a knowledge creating approach, *The TQM Magazine*, **12**, *2*, pp. 125-36.

Beecroft, G.D. (1996) Internal quality audits – obstacles or opportunities? *Training for Quality*, **4**, *3*, pp. 32-34.

Bessant, J. & Francis, D. (1999) Developing strategic continuous improvement capability, *International Journal of Operations & Production Management*, **19**, *11*, pp. 1106-19.

Bond, T.C. (1999) The role of performance measurement in continuous improvement, *International Journal of Operations & Production Management*, **19**, *12*, pp. 1318-34.

Bou-Raad, G. (2000) Internal auditors and a value added approach: the new business regime, *Managerial Auditing Journal*, **15**, *4*, pp. 182-86.

Dale, B. & Bunney, H. (1999) *Total Quality Management Blueprint*, Oxford: Blackwell Publishers.

Fellows, R. & Liu, A. (1997) *Research methods for construction*, Oxford: Blackwell Science.

Ghobadian, A. & Gallear, D. (1997) TQM and organisation size, *International Journal of Operations & Production Management*, **17**, *2*, pp. 121-63.

Gieskes, J.F.B. & Broeke, A.M. (2000) Infrastructure under construction: continuous improvement and learning in projects, *Integrated Manufacturing Systems*, **11**, 3, pp. 188-98.

Harrington, J.H. (1995) Continuous versus breakthrough improvement: Finding the right answer, *Business Process Re-engineering & Management Journal*, **1**, *3*, pp. 31-49.

Harrison, A. (2000) Continuous improvement: the trade-off between self-management and discipline, *Integrated Manufacturing Systems*, **11**, *3*, pp. 180-87.

Hellsten, U. & Klefsjö, B. (2000) TQM as a management system consisting of values, techniques and tools, *The TQM Magazine*, **12**, *4*, pp. 238-44.

Kaijser, L. & Öhlander, M. (1999) *Etnologiskt fältarbete*, Lund: Lund Institute of Technology (in Swedish).

Karapetrovic, S. & Willborn, W. (2000) Generic audit of management systems: fundamentals, *Managerial Auditing Journal*, **15**, *6*, pp. 279-94.

Kaye, M. & Anderson, R. (1999) Continuous improvement: the ten essential criteria, *International Journal of Quality & Reliability Management*, **16**, *5*, pp. 485-506.

Laszlo, G.P. (1999) Implementing a quality management program – three Cs of success: commitment, culture, cost, *The TQM Magazine*, **11**, *4*, pp. 231-37.

Love, P.E.D., Li, H., Irani, Z. & Holt, G. (2000) Re-thinking TQM: toward a framework for facilitating learning and change in construction organization, *The TQM Magazine*, **12**, *2*, pp. 107-17.

McAdam, R. (2000) Three leafed clover?: TQM, organisational excellence and business improvement, *The TQM Magazine*, **12**, *5*, pp. 314-20.

Michela, J.L., Jha, S. & Noori, H. (1996) The dynamics of continuous improvement: Aligning organizational attributes and activities for quality and productivity, *International Journal of Quality Science*, **1**, *1*, pp. 19-47.

Nicholson, N., Schuler, R.S. & Van de Ven, A.H. (1995) *The Blackwell Encyclopedic Dictionary of Organizational Behavior*, Oxford: Blackwell Publishers.

Orwig, R.A. & Brennan, L.L. (2000) An integrated view of project and quality management for project-based organisations, *International Journal of Quality & Reliability Management*, **17**, *4/5*, pp. 351-63.

Samuelsson, P. & Nilsson, L-E. (2002) Self-Assessment practices in large organisations: Experiences from using The EFQM Excellence Model, *International Journal of Quality and Reliability Management*, **19**, *2*, pp. 10-23.

Santos, A., Powell, J.A. & Formoso, C.T. (2000) Setting stretch targets for driving continuous improvement in construction: analysis of Brazilian and UK practices, *Work Study*, **49**, *2*, pp. 50-58.

Savolainen, T. (1999) Cycles of continuous improvement: Realizing competitive advantages through quality, *International Journal of Operations & Production Management*, **19**, *11*, pp. 1203-22.

Sommerville, J. (1994) Multivariate barriers to total quality management within the construction industry, *Total Quality Management*, **5**, *5*, pp. 289-98.

Sommerville, J. & Robertson, H.W. (2000) A scorecard approach to benchmarking for total quality construction, *International Journal of Quality & Reliability Management*, **17**, *4/5*, pp. 453-66.

Sun, H. (2000) Total quality management, ISO 9000 certification and performance improvement, *International Journal of Quality & Reliability Management*, **17**, *2*, pp. 168-79.

Tam, C.M., Deng, Z.M., Zeng, S.X. & Ho, C.S. (2000) Performance assessment scoring system of public housing construction for quality improvement in Hong Kong, *International Journal of Quality & Reliability Management*, **17**, *4/5*, pp. 467-78.

Yin, R.K. (1993) *Applications of a Case Study Research*, Newbury Park, USA: Sage Publications.

Chapter 19

Design Research and the Records of Architectural Design: Expanding the Foundations of Design Tool Development

Robert Fekete

Introduction

Considering the many mutual interests within the fields of design and architecture, it is surprising that so little architectural theory seems to have found its way into modern design research. Apart from a few exceptions, the abundant writings of architects do not add to design theory any more than in the form of casual guest appearances on the stage of design research, conforming to its problem domain and vocabulary, and therefore mostly dealing with questions concerning design activity. One cannot help wondering whether important aspects of architectural design treated in the rich sources of architectural theory have been overlooked during the last few decades of design research.

Has architectural theory really contributed to its fullest extent to the field of design research? Judging from the current state of affairs, the answer to this question seems to be an emphatic negative. The ultimate aim of the research project behind this contribution is to improve the modern design tools of architects, focusing on their functionality during the early stages of design. Based on the above assumption, a more open-minded and thorough review of architectural theory and the writings of architects in the context of design research seems to be one of the first steps in the right direction. This chapter argues that the findings of such a review could make significant contributions to the field of architectural design and the development of its modern tools.

State-of-the-art review

On the legitimacy of the writings of architects

A glance at the history of CAD research, and especially research in computer-aided architectural design (CAAD), reveals that the study of design activities of architects in practice and education has been a common approach towards building a foundation for the development of software. CAAD research in itself has

> 'extended and developed design research, which began in the 1960s. Indeed, the links between design methods and CAAD are so intertwined that it is almost impossible to separate them' (Tweed 2001).

Significant, however, of this research is the generalisation of architects into *typical groups*, represented by the ideal-type of designer devoid of any biographical history. The subjects of interviews, observational studies, statistical surveys etc. are mostly educational institutions, practices or some other notion of the profession at large. The sweeping statement that architects

> 'form homogeneous communities of likeminded people whose main activity is to design buildings' (Tweed 2001)

seems to be frequently made in CAAD research today. Hence the diversity of individual ways of design is effectively excluded. In the end, this can be seen as one of the factors behind the streamlined CAAD programs of today. This is where methods of design research could be complemented by architectural theory and the writings of individual architects in order to expand the foundations of CAAD development.

Apart from the generalisations being made in the studies of designers, the practice of predominantly studying design activity may be an insufficient approach in itself. In order to contribute further to CAAD development, architectural design research should make efforts to shift focus to *what* is being designed from *how* it is being designed. Archer (1979) supports this thought when saying:

> 'I never did like that hybrid expression *design methodology*. My objection was not only to the corrupt etymology, but also to the impression, conveyed by the term, that the student of design methods was exclusively concerned with procedure. For my own part, the motive for my entering the field ... was essentially ends-directed, not means-oriented'.

In the context of CAAD, O'Connell (1983) puts this clearly when declaring that

> 'the slow development of CAAD systems, indeed, is partly explained by the difficulty of deciding what is meant by architectural design'.

Considering the fact that CAAD development has so rarely turned to the field of architectural theory for input and ideas, the development of CAAD systems has, apart from being slow (which might have been more true in the 1980s), shown itself to be limited in its usefulness. In practice, today's tools conform to the duties of a relatively narrow (however numerous) segment of professionals, namely those involved in creating production drawings.

As a consequence of the limitations of CAAD, one could even argue that the architect's main tool of today not only excludes certain parts of the design process, it also contributes to an age-old form of segregation in the architect's offices: the one between the *designer* architect, whose main areas of work are excluded by CAAD, and the draughtsman whose tasks are promoted by it. True enough, the capabilities of modern software rarely benefit users as extensively as is technically possible, even during the stages when production drawings are being prepared or later in the process stretching into facilities management. Questions regarding CAAD use in practice today are partly the subject of this project's sister project represented by Chapter 20, *Communicating project concepts and creating decision support from CAAD*. However, to question whether the philosophical and theoretical foundations of today's CAAD programs are sufficient, or even correct, could well prove to be a fruitful way to widen their use into the early stages of design. As it is, these foundations seem to be all too rudimentary. The world of architectural theory, being the philosophy of architecture, may extend the vocabulary of CAAD programs and enrich the semantics of the objects and phenomena they are (in)capable of representing.

Architectural theory, that is herein mainly understood as the writings of architects, regularly deals with the results of design, i.e. architecture, rather than the nature of design activity. Complementing, for example, protocol analysis as a verbal account of what goes on in a designer's head during (or shortly after) design activity with likewise verbal, however written, accounts focusing on the design itself seems important. This chapter argues that design goals are best studied through the written records of architects; they are the shortest way to the designer's mind, providing a grip on not only facts, but also values. The explanation of this is quite simple. The nature of protocol analysis vitally differs from the situation in which architects reflect on their own work and try to put experience and theories into words, motivated by their own wish to do so. In the act of writing, the text becomes the main issue, a product of its own.

The study of design activity in a laboratory environment is associated with problems. For example, it is a necessity that the subject of the experiment is presented with a given design problem. However, this may yield some unwanted effects. One does not have to be a professional designer to sense the fact that the moment a specified design task is pointed out, a range of presuppositions have already been made about the goal of the design activity, and to a certain extent the course is already set out. For architects such definitions are usually subject to many considerations and re-considerations, personal as well as those emerging from contacts with clients and users, before being pinned down.

This process is all in the nature of dealing with architectural design problems. In experimental studies of design activity, these elements, while having major influences on the final result in real-life projects, seem to be excluded. The designer studied under these circumstances is presented with a goal and then asked: how would you go about reaching it? The obvious risk involved with a procedure where the problem, partly or wholly, is formulated beforehand is that it instantly produces a mental image, hence largely omitting the possibility for the designer to answer the crucial question: what is my goal?

The ambiguous nature of architectural design

Deciding what architectural design is, as O'Connell remarked, not an easy task. Upon presenting in a brief chapter a few architects' different views on architecture, Lawson (1990) concludes that:

> 'the intention [of the presentation] was simply to suggest that it is not necessary to include revolutionary or fringe ideas about architecture in order to find considerable variation in approach to the design process'.

Hence, ambiguities surround architectural design regardless of quality, scale, complexity and the like. Not surprisingly, difficulties are bound to occur when trying to establish a workable connection between an area as hard to define as architecture, and a computer program such as CAAD that requires clear-cut information in order to function. An attractive starting point in an attempt to achieve such a connection is to explore further the view that architectural design, especially in its early stages, mainly deals with ill-structured or wicked problems (Simon 1973; Rittel & Webber 1973). Since so many of the characteristics of wicked problems seem to apply to architectural design, especially during its early stages, a short review here seems worthwhile. Rittel & Webber propose ten aspects that characterise a wicked problem.

- There is no definitive formulation of a wicked problem.
- Wicked problems have no stopping rule.
- Solutions to wicked problems are not true or false, but good or bad.
- There is no immediate and no ultimate test of a solution to a wicked problem.
- Every solution to a wicked problem is a one-shot operation; because there is no opportunity to learn by trial-and-error, every attempt counts significantly.

- Wicked problems do not have an enumerable (or an exhaustively describable) set of potential solutions, nor is there a well-described set of permissible operations that may be incorporated into the plan.
- Every wicked problem is essentially unique.
- Every wicked problem can be considered to be a symptom of another problem.
- The existence of a discrepancy representing a wicked problem can be explained in numerous ways. The choice of explanation determines the nature of the problem's resolution.
- The planner has no right to be wrong.

A couple of these characteristics seem to relate immediately to what has been discussed before. Wicked problems would, for example, allow us to include the differing individual ways of reasoning that occur when it comes to architectural design, simply because one of the characteristics of a wicked problem is that there is no *one* correct solution to it. The outcome of a design process, for example a finished building, is judged differently by different parties involved, due to their individual interests and ideologies. Hence, a purely objective statement of its success is impossible. Rittel & Webber (1973) elaborate upon this by saying that:

> 'the formulation of a wicked problem is the problem' and 'the process of solving [it] is identical with the process of understanding its nature'.

Probably for many architects such a view is quite an appealing way of looking at things. Thomas & Carroll (1979) further add to it by saying that designing is essentially 'a way of looking at a problem', not exclusive to any particular type of problem, hereby leaving it to the problem-solver to decide how the task at hand should be treated and even whether it, in fact, should be perceived as an ill-defined or well-defined problem. This view is related to the one represented by Simon (1973), whose definition of ill-structured problems as a 'residual concept' gives a handy overall grip of their nature. Initially, he says, ill-structured problems are simply identified by what they are not. He continues by stating that there is no way of clearly classifying a given problem as either well-structured or ill-structured. In order for them to be manageable, he suggests that ill-defined problems are decomposed into sub-problems, hereby effectively turning them into well-defined ones, which in turn are subject to conventional problem-solving methods. Such a working method, abstractions and generalisations, is common to architects.

Largely, within the field of architectural design, ill-defined problems are fewer, more extensive and occur earlier in the design process, while well-defined ones, subject to more routine solutions, are many and scattered all over the later stages of design. This is a pattern quite typical for architectural design, unlike for example the design of cars, where first and foremost the chronology can be said to be more or less the reverse. Concerning architecture, this relates to another couple of characteristics of wicked problems, one being that every wicked problem is essentially unique. According to this definition, wicked problems cannot be classified, since there always might be important properties yet to be discovered that distinguish a specific problem from other seemingly similar ones. Hence to be manageable, wicked problems require complex judgements about their acceptable level of abstraction.

> 'Part of the art of dealing with wicked problems is the art of not knowing too early which type of solution to apply' (Rittel & Webber 1973).

Furthermore, stating that 'every wicked problem can be considered to be a symptom of another problem' is not just an analogy of the old saying that one should strive to get to the root of a problem. Typically, a wicked problem is multi-layered; in solving one aspect of it another more complex problem may reveal itself. However, as Simon pointed out, in practice one has to work on a level where things are manageable. The difficulty, though, lies in knowing when, and into what, to break down any given ill-defined problem, since every such decision significantly guides the course of the design process. Too high a level can render it unmanageable, while the contrary can result in solutions that only serve to reinforce higher-level problems, hence making things worse. In architectural design such decisions, and eventually the decision to end one's work on a particular design as a whole – to stop designing – is guided by external considerations, such as sufficiency, limitations or feeling; for example, 'this works well enough', 'we have exceeded the budget' or 'I like it the way it is'. Hence, 'wicked problems have no stopping rule'.

Seemingly one could delve deeply into the relevance of wicked problems for architectural design. For the time being, however, it is probably safe to conclude that the concept of wicked problems makes up a convenient container for analysis somewhere in-between the ambiguities of architecture and the formalised world of computers. It makes things graspable without being able (or wishing) to pin them down; hence, it relates to both areas. Possibly the concept of wicked problems could work as a filter through which architectural theory is poured for computer-friendly formalisation.

The shortcomings of today's CAAD programs

In Rittel & Webber's (1973) list characterising wicked problems, the last paragraph, 'the planner has no right to be wrong', may sound somewhat alarming. The authors seem to mean that in practice the principle of conjecture and refutation is not acceptable when it comes to wicked problems, thereby contrasting the world of science. Since the implemented solutions of wicked problems are effectively irreversible and as such can have great consequences for generations to come, planners cannot afford to make serious mistakes. In the case of buildings this mainly applies to the great costs involved in correcting mistakes; in the case of large-scale infrastructure projects, on the other hand, the consequences might well be truly irreversible. This is the context in which CAD programs were meant to make a considerable difference, as an aid for designers, in the end being the ultimate quality assurance tool for the construction sector. The prospects of CAD being able to achieve such a position still seem promising. However, in the context of architectural design today, particularly during its early stages, the benefits of CAD are far from being that significant.

It is important to remember that design research before the CAD breakthrough was dealing mainly with the improvement of design methods. Not much discussion concerning the tools of the designer was ever deemed necessary until the introduction of CAD. Design theory, as the basis for design tool development, is a result of the problems surrounding this new tool. Where design tools once were of more or less common nature in different design areas (pen and paper, physical models), various disciplines have developed specialised software to suit their specific needs. Within the construction process, CAD for architectural design (CAAD) has taken a course of its own.

The main role of CAAD programs in architectural practice during the last 10 years has been that of the support tool in the later stages of the design process, the advantages of which today seem quite clear. In contrast, the benefits of using CAAD during the early stages of design are uncertain. Indicative of this is the fact that application of CAAD early on in a project today, for example, for sketching, is seldom done. The main objections on the designer's part towards such practice concerns the difficulties associated with modelling the loosely determined information so typical of the early stages of design. This is coupled with the view that it is unprofitable having to specify information more than necessary, simply for software handling reasons. Even when the initial wish in a project is often simply to visualise ideas concerning volume and functionality, the cumbersome input of information constantly causes the designer to lose his or her train of thought, and the feedback is poor. As a result, the use of today's CAAD programs in the early phases of design is judged simply not worthwhile. Hence, an important question to answer seems to be:

> *What are the actual benefits of computerising architectural design information in the early stages of design?*

Ekholm and Fridqvist, creators of the BAS•CAAD project, state that

> 'today's CAD software cannot provide design evaluation tools until later

stages of the design process when much of the design has become fixed' (Fridqvist & Ekholm 1996).

Consequently,

> 'the interpretation of the acronym CAD as computer aided design is questionable in the realm of building design and the tools available today' (Fridqvist 2000)

meaning that CAD should in fact spell computer aided draughting. The popular argument from the side of the proponents for the introduction of computer support early in the design process emphasises the advantages that digital information brings to the later stages of the process.

> 'It has long been held by researchers that computer aided design needs to be continuously used from the earliest design phases, if the most benefits are to be gained' (Fridqvist 2000).

Though again, due to the insufficient functionality of the software in the early design stages:

> 'a considerable amount of the information gathered during building design today is not forwarded to the subsequent stages of the construction process' (Fridqvist 2000).

Important issues related to this, and hence the overall functionality of CAAD programs, are partly those concerning modelling in itself, or information input (i.e. *what* and *how*), and partly those relating to communication or information output (i.e. to *whom* and *when*). Consequently, following the previous question posed, the next question to answer would be:

> *What properties should a tool for computerisation of architectural design information have in order to be of benefit to architects for use in the early stages of design?*

The client's evaluation and categorisation of architects and their practices today have come to include a whole new criterion, namely the status of the architect's equipment. Historically, this has never been regarded as indicative of the architect's competence. It seems unacceptable that the great investments made today in architect's offices in hardware and software, constantly craving maintenance and upgrading, should be induced by competition only and/or a mere technology push. The paradox between the use of a tool being obligatory in order to uphold one's professionalism and this tool's inability to be a true asset in the profession is unique in the design sector in many ways. Recognising the fact that computer aided support for architects in the early stages of design is highly insufficient, and assuming that one of the major reasons for this is that architectural theory and the writings of architects have been largely overlooked within the realm of CAAD research, this research project will initially deal with the following questions:

- What contributions can architectural theory and the writings of architects make to design research?
- How can architectural theory and the writings of architects expand the foundations of design tool development?

One of the reasons for the shortcomings of CAAD, indicated above, are the methods used when studying the design activities of architects, methods characterised by Archer as being

> 'the product of an alien mode of reasoning' (Archer 1979)

inevitably leading to some sort of generalisation of the software produced. One might argue that the view represented by these methods contradicts the nature of the early design stages, in that design activity early on mainly concentrates on identifying the design problem, resulting in individual solution concepts, that in turn are in need of varying kinds of support for their representation.

The prospects for CAAD programs

Continuing his thought, as quoted in the beginning of this chapter, Archer (1979) states that by getting involved in design research:

> 'I was concerned to find ways of ensuring that the predominantly qualitative considerations such as comfort and convenience, ethics and beauty, should be as carefully taken into account and as doggedly defensible under attack as predominantly quantitative considerations such as strength, cost, and durability'.

He continues:

> 'My present belief ... is that there exists a designerly way of thinking and communicating that is both different from scientific and scholarly ways of thinking and communicating, and as powerful as scientific and scholarly methods of enquiry, when applied to its own kinds of problems'

i.e. ill-defined problems.

The philosophical foundation of CAAD programs of today is the rationalistic view that the design tool should be able to *represent* the conceived reality (Turk 2001). Striving to model *fully in 3D* has become equal to the quest to achieve as detailed a representation of the finished building as possible. Consequently, to extend the amount of *symbols* that represent real life objects and their properties has been considered to be largely the correct path to follow in the development of today's software. The grouping of these objects into object-classes is what mainly constitutes the basis for development towards so-called object-oriented CAAD programs. Taking the above discussion about architecture being an ill-defined problem into account, many difficulties arise around such a narrow-minded course of development. Apart from the fact that ready-to-use symbols have to be preconceived for objects not yet thought of, let alone defined, a preconception of an entire domain, i.e. architecture, becomes inevitable. The generalisation of architects into groups handy for the purpose of study has also been mentioned above.

An interesting parallel to this is the simplistic way in which predefined object-classes have been identified in today's CAAD programs. Needless to say, the amount of predefined classes would have to be close to infinite in order to form suitable 'containers' for every conceivable object one would want to model. Nevertheless, such a division is the common structure of today's CAAD programs. Once again a sweeping statement has been made that walls, floors, roofs, pillars, slabs, doors, windows and various installations and furniture (sum total) are sufficient artefacts to model every conceivable physical part of a building. Forms and shapes not covered by these classes have to be geometrically modelled by the user, resulting in objects with little or no possibility for property attribution; in other words, they are destined to be 'unintelligent'. Furthermore, once the class belonging of an object is defined, it cannot be altered without the redefinition of the entire object itself. Consequently, as it is today, CAAD programs are neither very allowing, nor forgiving.

Concerning this last issue, recent research has taken steps towards more flexible object-oriented modelling. It has been recognised within the BAS•CAAD project, as well as elsewhere, that central to building product modelling is the abandonment of the traditional class-centred approach in favour of an object-centred approach (Ekholm & Fridqvist 1998). In the BAS•CAAD system this is handled by the concept of *schema evolution* or the *dynamic definition of design object classes*. Users develop their own project-specific *model schemes* containing object classes created with the support of predefined libraries, but nevertheless in essence freely defined by the users themselves.

To put it simply, not only should the creation of certain objects be the prerogative of the designer, but also the creation of the classes to which they are to belong. Conforming to the fact that the creation of classes can be a very demanding task, the libraries based on established building classification systems are supplied – a professional approach very much different from the *Lego*-philosophy employed today. A conclusion made in the BAS•CAAD project regarding the flexibility of objects is that the software has to, during the entire design process, enable the user to:

(1) create and alter classes suitable for the objects modelled; and
(2) freely redefine the classification of the objects modelled.

So far the discussion above has been concerned with the modelling of physical objects alone. However, architects are typically concerned with the non-physical sides of life as well. This quote, characteristically poetic, by Kahn (1964) is a good example of the complex reasoning often found in the writings of architects.

> 'The order of making the wall brought about
> an order of wall making which included the opening.
> Then came the column,
> which was an automatic kind of order,
> making that which was opening,
> and that which was not opening.
> A rhythm of openings was then decided by the wall itself,
> which was then no longer a wall,
> but a series of columns and openings.'
> (Kahn 1964)

To study the evolution of concepts, such as this one, is intriguing, since it becomes evident that many things, apart from hands-on physical circumstances, influence the way that things are finally perceived. When initially Norberg-Schulz (1976) clarifies his view on the differences between space and place, he uses a common expression deriving from our everyday use of language. In short, according to him, place is space with identity, and by emphasising that events *take place* in a space, he highlights perhaps the most important of its many identities, namely its user's activities. In such a seemingly simple manoeuvre, Norberg-Schulz turns the abstract concept of space into a more concrete notion of place, hereby isolating one aspect of place that in his view distinguishes it from its literally more anonymous predecessor.

From a scientific viewpoint, Norberg-Schulz might not seem to have a very strong chain of arguments (which is mostly to blame on the extremely shortened review of this particular train of thought in here). For a 'designerly' mind however, such definitions often represent a key for the future work of the designer. The importance of these *unscientific* concepts is therefore evident. (As a matter of fact few ideologies of architects resemble regular scientific systems, nor was this ever the intention of their authors. Kahn's words in particular are often of the philosophical kind, sometimes composed as laconic sentences with a poetic and mysterious ring to them, while at other times, especially during interviews, his answers to questions are more reminiscent of soliloquies.)

So far, today, the importance of being able to model quantitative data, such as for calculation, has been recognised, but should qualitative concepts such as identity also be modelled or in any other way supported by CAAD? The specific example of user activities, particularly in relation to building briefs, is one of the cornerstones of the BAS•CAAD project (Fridqvist 2000). However, the point made in this context is that this example represents only one of many concepts within the field of architecture for which the possibilities of manipulation with today's CAAD programs are poor. Basing his arguments on the phenomenology of the German philosopher Martin Heidegger, Turk (2001) answers the question above with ease. He justifies the coexistence of intangible objects alongside physical ones simply by stating that:

> '...we can think of them and that should, according to metaphysics, suffice. They are constructed and found useful in the skill – the "techne" – of the architects'.

One of the interesting elements in Turk's theories relevant for this discussion is his division of models into abstract models and concrete models. An ideal CAAD program would by this terminology be representing the abstract (or conceptual) model and, thus, would have the ability to house all of the different concepts of architecture, the ones existing as well as the ones yet to be invented. Since today's object-oriented CAAD programs can be said to represent the category of concrete models rather than abstract ones, it is not hard to understand why users so often run into difficulties when working with them. Turk gives an illuminating example of this. A concrete model would have its counterpart in something tangible, for example the Eiffel Tower, while abstract models denote the concept of towers in general. One can easily realise that it is relatively simple to produce a specialised CAAD program for designing *the* Eiffel Tower, because in effect the software would not have to contain any other objects than the ones forming this specific tower, but it is considerably more difficult to produce a program that assists in the building of every conceivable tower, let alone every conceivable building or, in the end, a CAAD program that would be suitable for every architectural task.

According to Turk (2001) such difficulties will prevail, as long as the development of CAAD is fundamentally based on today's object-orientated approach to modelling. He emphasises that:

> 'objects are [mental] constructs, inventions, not something that has existence independent of ourselves',

hereby effectively levelling all concepts deriving from the architect's mind and stating that they deserve equal support by a modern CAAD tool, irrespective of whether they are concrete, tangible or quantifiable or not. As it is today, he claims:

> 'the term conceptual [abstract] model for creative design is a contradiction in terms'

when resulting in software only suitable for supporting

> 'routine, bureaucratic processes that fit into the schema designed during software development. [...] Only designs that fit into the model are possible. Creative ones, by definition, are those that step out of the predefined conceptual model' (Turk 2001).

In the end, he says, model-based software that truly supports creative design would require the additional skills of a computer programmer to use, hereby criticising the kind of extensible models investigated in projects like BAS•CAAD, finally landing on the somewhat sullen note that still, today, the only tool that does not stand in the way of creativity is the pencil (Turk 2001). Being involved in the field of CAAD research, however, Turk believes that things could change, but as to what extent he will actually be able to present a working alternative to object-oriented modelling, or in fact *computer modelling* on the whole, remains to be seen. Still, his arguments forming a radical approach by eliminating objectification altogether, give the impression of being a welcome starting point from where to concentrate on defining the domain of architectural design in relation to CAAD, and to investigate further how software flexible enough to be able to adapt to every single change in this domain, or in other words, every single design project, should be conceived.

Archer (1979) brings us back to safer ground with a quote containing many key issues still valid for future consideration:

> 'Indeed, we believe that human beings have an innate capacity for cognitive modelling, and its expression through sketching, drawing, construction, acting out, and so on, that is fundamental to thought and reasoning as is the human capacity for language. Thus design activity is not only a distinctive process, comparable with but different from scientific and scholarly processes, but also operates through a medium, called modelling, that is comparable with but different from language and notation'.

With today's software it seems that if flexible enough modelling is to take place, especially during the early stages of design, we are restricted to manipulating pure geometry. As just one consequence, all of the non-physical concepts referred to in architectural theory will be excluded. Hence, we seem to be left with two choices. Either we settle for a pen-and-paper mimicking 2D software and free-form modellers imitating the likes of traditional cardboard model building (the value of which is questionable), or we skip CAAD support entirely in the early stages of design, and restrict ourselves to making use of CAAD once the objects that are to be modelled are well-known and defined.

One way to move forward, proposed by this chapter, would be to conduct an initial extensive survey of the different opinions of architecture and architectural design found in the sources of its theory, in order to (re)define exactly *what it is* we want to be able to use our design tool for. Today, the wish list for such things is already long, ranging from quite simple desires such as the ready ability to calculate quantities, to the more complex issues regarding the nature of architectural design. Probably that list is what initially has to be extended and rooted in architectural theory, and probably it should not simply result in a further expansion of the number of objects, or types thereof, handled by the software. The concept of object-classes (or even objects) in itself seems to be an all too restricting system to conveniently accommodate the phenomena and the rich knowledge constituting the wicked problem of architecture. In revising such a system, the logical next step might be to regard the design of a CAAD program as a wicked problem in itself.

Conclusions

This chapter has tried to point out that the finished result of architectural design is to a large extent determined early in the design process and that the traces of these decisions are to be found in architectural theory and the writings of architects. Since the shortcomings of today's CAAD programs are by far most significant during these critical early phases of design, the conventional methods of design research that have so far largely formed the basis of CAAD development need to be complemented. It seems important for architectural design research to shift focus to *what* is being designed from *how* it is being designed, in order to contribute further to the development of CAAD programs that could be of true benefit to designers. One way of initiating such a shift would be the merging of design research with architectural theory and the writings of architects.

References

Archer, L.B. (1979) Whatever Became of Design Methodology? In: Cross, N. (ed.) (1984) *Developments in Design Methodology*, New York: John Wiley & Sons, p. 347.

Ekholm, A. & Fridqvist, S. (1998) A Dynamic Information System for Design Applied to the Construction Context, *Proceedings of CIB W78 workshop The Life-Cycle of Construction IT* (ed. Björk, B.C. & Jägbeck, A.), Stockholm: Royal Institute of Technology, pp. 219–32.

Fridqvist, S. (2000) *Property-Oriented Information Systems for Design*, Doctoral thesis, Lund: Lund Institute of Technology.

Fridqvist, S. & Ekholm, A. (1996) Basic Object Structure for Computer Aided Modelling in Building Design, Conference paper presented at the CIB W78 workshop *Construction on the Information Highway*, Bled, Proc. Turk, Z. (ed.) Ljubljana: University of Ljubljana, pp. 197–206.

Kahn, L.I. (1964) Talks with Students. In: Latour, A. (ed.) (1991), *Louis I. Kahn, Writings, Lectures, Interviews*, New York: Rizzoli International Publications, p. 157.

Lawson, B. (1990) *How Designers Think*, London: Butterworth Architecture, pp. 121–33.

Norberg-Schulz, C. (1976) The phenomenon of place, *Architectural Association Quarterly 8*. in: *Arkitekturteorier, Skriftserien Kairos, Nummer 5 (1999)*, Stockholm: Raster Förlag, pp. 89–115.

O'Connell, D. (1983) An Educational Strategy for CAAD and its Implementation in a New System with a Sophisticated Interface, *Proceedings of the International Conference eCAADe Brussels (Belgium) 1983*, pp. I.1–I.19.

Rittel, H.W.J. & Webber, M.M. (1973) Planning Problems are Wicked Problems. In: Cross, N. (ed.) (1984) *Developments in Design Methodology*, New York: John Wiley & Sons, pp. 135–44.

Simon, H.A. (1973) The Structure of Ill-structured Problems. In: Cross, N. (ed.) (1984) *Developments in Design Methodology*, New York: John Wiley & Sons, pp. 145–66.

Thomas, J.C., & Carroll, J.M. (1979) The Psychological Study of Design. In: Cross, N. (ed.) (1984), *Developments in Design Methodology*, New York: John Wiley & Sons, pp. 221–35.

Turk, Z. (2001) Phenomenological Foundations of Conceptual Product Modelling in AEC, *International Journal of AI in Engineering*, **15**, pp. 83–92.

Tweed, C. (2001) The social context of CAAD in practice, *Automation in Construction*, **10**, pp. 617–29.

Chapter 20

Communicating Project Concepts and Creating Decision Support from CAAD

Jan Henrichsén

Introduction

In today's building design process, architects can decide to use object-oriented, solid modelling CAAD programs instead of 2D CAD and word processing. By using a modelling CAAD program, a digital model of a building can be built, carrying all information about it in order to perform simulations, undertake presentations and generate descriptions. Architects, as well as other consultants, clients and authorities, are not fully aware of the full benefit of using CAAD tools, and so the concept of CAAD is only partly implemented. Architects must, therefore, be convinced that a CAAD-derived model can be the core for work of the whole design process and that it can constitute the sole database from which it is possible to export all necessary documents.

The research forming the subject for this chapter involves a project in which a complete CAAD model of a building – in this case a multi-family apartment building – is going to be built, using a commercial program. The project is intended to show that it is possible to put all the information that is needed in the design process into a digital model and then export it to all common, more or less standardised documents.

State-of-the-art review

Nowadays, architects have economical and practical opportunities to work with computer programs foreseen since the 1960s, but accessible only of late. It is interesting to note that relatively very few architects actually try to use these kinds of programs. The author has worked as a practising architect since 1979 and has developed his personal professional daily work process from traditional hand drawing with pen on paper, via draughting CAD (2D) to working with object-oriented, solid modelling CAAD programs dreamed about in the 1960s. The following sections provide a brief overview of ideas and thinking about CAD techniques, counterbalanced by the author's personal experiences in this field.

CAAD ideas and thinking

What are the different computer methods that an architect can use today in the building design process? There are two main categories to choose from when contemplating CAD activities; first, using the computer as a draughting and modelling system; or second, using it as a design medium (Gero 1985 p.107). The different aspects of using CAAD as a design medium are covered by many researchers and are not the primary concern of this research. Instead, the modelling aspects in the phases following the initial design stage are of interest. The use of computer aids in the early stages of the design process is, of course, of immense interest to many architects and important to develop; but the use of CAAD in the latter stages of the process is as important when it comes to fulfilment of the design process. O'Connell (1983) states that:

> 'The slow development of CAAD systems ... is partly explained by the difficulty of deciding what is meant by architectural design'.

This helps to explain why there have been so many research projects over the years concerning architectural design and computer aids focused on the early stages of the design process. At the same time, commercial CAAD programs have developed in terms of producing documents for the latter stages to a level where the use of a CAAD program instead of a 2D draughting program can now be justified. A change from 2D draughting to CAAD modelling can now be implemented into the routine work of practising architects simply by purchasing commercial software.

At this point, it is probably useful to provide explanations of some of the terms used – 2D CAD, 3D CAD, object-oriented, solid modelling, computer-aided architectural design – or, rather, the important differences between them. Much has been said about these terms over the years and many statements are common. Penttilä (1989) is typical of those who strive to explain the different computer-aided design techniques:

> '2-D Draughting CAD systems offer tools to create 2-dimensional drawings with basic graphic primitives such as lines, arcs, curves or text-elements. [...] Draughting systems are mainly used to produce working drawings, substituting hand drawing, since their features seem to match best into this work. [...] A model created with a common 3-D CAD system is purely a geometric model with information about geometric basic primitives – just symbolically representing building components or concepts.'

The words of Björk (1995), regarding building product data models, can be used to comprehend the term solid modelling object oriented computer-aided architectural design:

> 'A building product data model models the spaces and physical components of a building directly and not indirectly by modelling the information content of traditional documents used for building descriptions.'

In his foreword to the book *CAD Principles for Architectural Design* by Peter Szalapaj, one of the pioneers of CAAD, Aart Bijl argues that:

> 'Developments of CAD in architecture have a somewhat back-to-front history. Early developments during the 1960s and 1970s included computer modelling of architectural forms, which supported a range of analyses to do with building performance and buildability ... Later, during the 1980s and 1990s, as computers became smaller and less costly, so new developments had to be targeted at whole markets, such as the market consisting of all architects. Market economics also resulted in computer programmers becoming very distant from end users. Consequently developments became less ambitious, offering drawing and rendering systems that carried no data about what was being depicted. It has taken a long time to get back to design modelling and analysis' (Szalapaj 2001).

Even if you can trace ideas about CAAD back to the 1960s, you would find that very few architects within architectural design offices actually had access to computers (and CAAD tools) before the middle of the 1980s (Fridqvist 2000). When the CAAD era finally dawned for those architects, it was pure 2D draughting tools that occupied most of their CAD time. Eastman (1999) states that 2D systems best responded to the requested task of producing production drawings and that the dominant usage still was based on using CAD as a graphics editor. Eastman remarks that:

> 'The challenge before us is to develop an electronic representation of a building, in a form capable of supporting all major activities throughout the building life cycle.'

He also offers the opinion that it is possible that the user community's recognition of the advantages of building modelling could evolve only after a period of electronic drawing and simple add-on applications.

Without starting a discussion on the impact of the marketing power of certain CAD programs (or rather their vendors) we can stress that instead of modelling tools, it was draughting tools that were increasing in number in architectural design offices. This was the case despite statements of the form:

> 'Already today's tools will allow digitising of whatsoever data, and this will also be the case in the future...' (Penttilä 1989).

Nonetheless, a few architects involved in practical (as distinct from research and development) projects used commercial modelling programs and made solid building models, transferring data and plotting drawings from the model core. At the time, they lacked theoretical knowledge of object-oriented, solid modelling and they did not appreciate exactly how the programs worked deep inside, but they used them daily. Therefore, it is interesting to read literature about object modelling from the years around 1990, where the authors still talk about prototypes and mostly discuss different data transfer formats and the standards that modelling programs should follow (Björk 1993; Wikforss 1993).

While practising architects in those years successfully communicated (and still do) with other consultants using the two AutoDesk (the software company developing AutoCAD) file formats DXF (Data eXchange Format) and DWG (AutoCAD's internal format), progress in the standardisation field covered file transfer formats like STEP (Standard for the Exchange of Product Model Data, which is an ISO standard); NICC (Neutral Intelligent CAD Communication); and lately IFC (Industry Foundation Classes). Few examples are to be found of the latter's use by practising architects.

Even today, we see few practical examples of the creation of a robust CAAD model, with all the information about the building within the database. True enough, there are some architects working with modelling tools and manipulating different information from the database, enabling them to follow their own direction of interest, but no one seems to have built *the* model, using all the possibilities available in a CAAD modelling system.

Examples of CAAD model building

Much of this section is based on observations by the author, reflecting his experiences in practice over some 13 years. During this period, many building models were created that can be seen as examples of different methods of building data models, from the first struggling attempts to more refined and confident methods. Many were built as commercial models in an architectural design office, in collaboration with clients, authorities and other consultants. Models were also built in collaboration with researchers at Lund University. The CAAD program being used was the commercially available CAAD software, *ArchiCAD*, by Graphisoft in Hungary: a popular package in Scandinavia and the UK.

The first tentative models built in the office in the late 1980s were merely meant to be used as plan drawings, mostly at a scale of 1:50. For presentational purposes, the 3D information was also used to some extent. From the very first project, digital communication with other consultants was established, but the communication handled only plan drawing information. Instead of sending transparent drawing copies, DXF files were now sent on floppy disks. There were some technical/practical problems in the beginning but they were solved during the process. The various actors learned quickly, and after some projects all kinds of plan drawings to different scales were exported from the models, along with sections and façades. Although at that time it was possible to add more information (costs, manufacturer, delivery times and so on) to the different objects it was never done. Clients did not ask for it and they were unwilling to pay for the extra information. The full database information in the models was not used to any true extent, mostly because of a lack of time for active experimentation in the late 1980s.

The following years saw efforts in CAD modelling increase through the more intelligent use of the system. For each new project, the layer system became more and more important and was progressively developed. Finally, the system had almost the same structure as the *de facto* industry standard, BSAB (Byggandets Samordning AB) system, but with other names and numbers. Consequently, the homemade layering system was changed to a layout following that of BSAB. But since BSAB only covered the actual building and not any parts of the design process, some extra layers were added to the system concerning pure 2D objects such as texts, lines and so on. Some other layers were also added for specific 3D perspective reasons. *ArchiCAD* handles some building parts in the way that they reside in a specific library and pointers are placed from the model to those objects. This library was also organised following the BSAB system, as catalogue libraries are normally organised in architectural design offices.

Soon afterwards, at Lund Institute of Technology, it was possible to test different methods in building solid models. This began in 1991 at the Department of Computer Aided Architectural Design and the Department of Construction Management, and several models followed. At the same time, a quantity and cost calculation program, *SYRE*, was developed. *SYRE* could import values from the database of a CAAD model, enabling cost and time data to be linked; however, it was not necessary to add these data to the CAD model's database. *SYRE* combined the exported quantities with cost and time data allowing cost and time estimates to be produced. This feature was used subsequently in practically all models developed in Lund.

The models generally adopted different approaches to the treatment and use of a solid model. The first model was used to make a comparison between two types of information – in this case a 3D model versus plain drawings – to determine their comparative usefulness in supporting decision-making concerning the building. One model demonstrated alternative solutions. Another model illustrated the working site and the construction work. By using the different drawings from the designers and the timetable for the construction, the model was built in chronological order, week-by-week, to produce what is commonly termed a 4D model – see Chapter 21, *Using 4D CAD in the design and management of vertical extensions to existing buildings*. This model was also used for the working site planning. One model, consisting of many, almost identical, buildings (in this case, houses), was used for quantity/cost estimation and presentation to the client, users, authorities and the public.

Economic and technical comparisons between a timber-framed structure and the same building with a concrete structure formed the target application for one of the models. The aim of the latest model-building project in Lund is to understand, from the early design stage, the ways in which buildings can harm the environment. In this case, environmental data were not added directly to the model database. Instead, they were exported to *SYRE* in the same way as mentioned earlier for cost and time.

Research project

Project description and objectives

One of the primary benefits from the experience gained in the modelling activities and developments described in the previous section was that a variety of data could be introduced into the database model of the building. This has opened up many possibilities that are now being investigated as part of the author's research project. The main objective of the project is to demonstrate the practical usefulness of an object-oriented, solid modelling CAAD program in the design process, the construction/production process, the facility management process and the demolition process. The expectation is that the CAAD modelling tool will, if it is used in an *intelligent* manner, create cost estimates, check the building's impact on the environment and produce schedules for on-site working; other applications are possible. Another objective of the research is to show how a CAAD program used this way can enhance communication between the different actors in the design and construction process and double as a decision support system for the architect and others with specific responsibilities.

The initial steps are to show the extent of what is possible so that fruitful lines of research and development can be pursued. One of the early tasks is to secure end-user interest by demonstrating that it is possible to produce traditional documentation, to satisfy the needs of the different actors and project stakeholders, from the database of a CAAD model. Thus, the first phase of this project will concentrate on creating an object-oriented model using a typical, PC-based commercial CAAD program and, from the database, export traditional documentation needed in the construction process.

Research methodology

The overall approach to the first phase of the project relies on a modified model of problem solving advanced by Bunge (1983) – see Fig. 20.1.

The *problem* is that architects continue to produce standard documentation using 2D draughting programs. The *hypothesis* is that it is possible to produce the same standard documents from the database of a CAAD model. The *empirical survey* will be to build a solid CAAD model and export the documents. The *theoretical survey* will be to summarise building and documentation standards, as used in Sweden, and to import them into the structure of the model. The *control* will be the comparison between the documentation exported from the model and that prepared in a traditional way. The *evaluation of the results* could suggest that the CAAD model should be further tested for new benefits in the design process.

Standard documentation

Before making the model and exporting documentation from it, it has to be clear what documents are to be expected. Among designers there is consensus as to what are standard or normal documents in the design process. Standard and normal are, of course, words that architects normally choose not to use and they will perhaps be reluctant to admit that they produce standard documents. It is natural that all offices do not produce exactly the same documents. Nevertheless, there are similarities in the documents produced by different

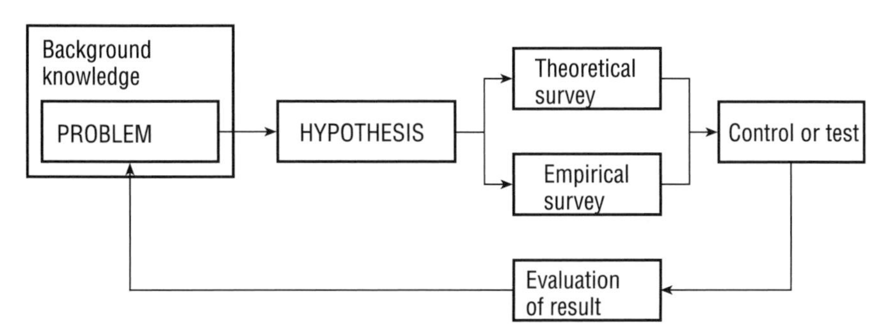

Fig. 20.1 Description of the problem solving nature of the first phase of the research, after Bunge (1983).

offices and if one kind of standard document can be exported from the database, certainly other, similar documents can be produced as well.

As implied above, knowing exactly which are the standard documents and which are not is not of particular importance. Some kind of confirmation as to what standard documents are has, however, to be accomplished in collaboration with other actors in the design process. One method of finding out which are the standard documents could be to ask a number of architectural design offices and others; for example, authorities and contractors, to present their documents from different projects, compare them and then finally settle on some notion of the kind of standard documents used by a corpus of offices.

If the main purpose of this project was to be about identifying standard documentation, it would be necessary to undertake a thorough survey. Since the main interest in the project is to show that it is possible to export documents, preferably standard documents, from the CAAD model, a simplified method is sufficient; the documents have to be plausible as far as being representative of what is standard. The method will involve producing and presenting a list of suggested documents to an architectural design firm, a public authority and a contractor. After discussion with these collaborators, it is expected that a good enough list of plausible documentation will have been produced.

CAAD model

Once the kinds of documents that are needed from the architect have been determined, it is then possible to proceed to model building. First, some words about this procedure. If the purpose is to build a model only for exterior studies of volumes and so on, it is a rather simple task. In that case, the model is merely treated in the same way as a clay model, though a little more intelligent; but it is still, more or less, like a simple clay model. If the model is intended to be used for purposes besides volume studies, it is more complicated to build, or more correctly, to structure the model. A very strict scheme has to be followed when the model is built, if the intention is to put all possible information into it and to export all types of documents, including drawings, from it.

This method of model building is leading up to what, in this chapter, is meant by a CAAD model. A complete, real building is built, but digitally. Every element in the model of the building must carry the same information as the element in the real building. If a certain piece of timber in the real house is made of pine, has the dimensions 45 × 45 × 1000 mm and is painted white in three layers, using a specific paint and so on, the correspondent digital part in the CAAD model must have the same information in the database. Otherwise, it would not be possible to present all the information about it in the required documents.

In order to be able to represent specific building parts only in certain documents, one also has to create a system that directs the representation of the parts to the correct drawings and text documents. This can be done with the help of layers and layer combinations. The layers correspond to different building classification codes and the different parts are put into corresponding layers. Layer combinations gather parts belonging to the same specific representation. Instead of using layers, the specific building classification codes can be connected directly to the parts themselves. This latter method is certainly the more correct way of building a model, as each part should carry all necessary information, including the code needed by the classification system.

The model will be based on *ArchiCAD* 7 from Graphisoft, and is a standard product. For this reason, some localisation is needed. Preferences need to be set to cater for different regulatory requirements, customs and practices. However, experience of the use of this particular software package means that some aspects can be shortcut. Settings to change are, for instance, the layering, the object library organisation and drawing frames. Since the model is going to take benefit from the Swedish standardisation systems, the layer system has to be structured following the BSAB system. Likewise the object library has to be organised in the way the BSAB is structured.

The actual model in this project will be one of a fairly simple building, but one that is fully designed though not necessarily built. The intention is to use the full documentation for a building, designed by an architectural design firm. The model will be built thoroughly following the design details produced by the architect.

Exporting documents from the CAAD model

When the model is fully built, the next step will be to export the different documents from the database. *ArchiCAD* has a number of templates and other functions to create documents in addition to drawings. That said, drawings have never been a problem to produce from a solid model and in this research project the drawings will, of course, be shown too. The intention is to export and print or plot the different documents, so that a complete set of documentation concerning the building is created. These documents will then be compared with the original set of documents produced by the architect of the building from which the model was built. The reason for the comparison is to see whether or not the documents made in these two different ways communicate the same information.

Research results and industrial impact

Quantification of results

Besides the solid model built in this project, the results are primarily the expected standard documents. So far, it has not been decided which documents will be produced. A tentative proposal is as follows:

- plans, façades and sections at different scales;
- elevations at different scales;
- details at different scales;
- materials and workmanship specification;
- room descriptions;
- quantity calculation (export to *SYRE*);
- still exterior images (perspectives); and
- still interior images (perspectives);

plus the following documents:

- walkthrough animations;
- VR animations; and
- shadow pattern animations.

Implementation and exploitation

It is inviting to believe that architects will, in the near future, use CAAD models as the major tool in the design process. The expectation is that this project will have contributed to the implementation of CAAD modelling methods. In the long term, this first step will have hopefully led to the full implementation of CAAD modelling in which all necessary information used in the project from design, through construction, to facilities management and demolition, will be integrated. This hope can perhaps be considered as a true form of building simulation. That said, the research project does not intend that new CAAD programs should be created, but that the full potential of suitable commercial CAAD modelling programs may be demonstrated. In this simple but direct way, it is hoped that designers and other actors will move closer together to accomplish a higher level of communication that will serve the client.

Conclusions

The project discussed in this chapter is intended to show that it is possible to put necessary information into the database of a CAAD model, and thereafter extract it to produce traditional documents. By building a complete CAAD model using a normal commercial CAAD tool and presenting all standard documents expected in the building design process, this project can perhaps encourage architects to move from the use of 2D draughting systems to modelling tools. A first step is to help architects become confident in their use of modelling programs by demonstrating that they are capable of producing documentation in forms that are consistent with their needs and, importantly, those of the other actors and stakeholders in a project. If this is achieved, one step has hopefully been taken towards a more refined use of CAAD modelling tools. One of the aspirations for this focus on a more dimensionally complete approach to handling design and other information is that it may help the different actors in the process to communicate their ideas better through a common or shared model. It will reduce errors and provide a consistency of purpose in driving through an evolving and better-elaborated model of the building, upon which owners and occupants can place real value.

References

Björk, B.-C. (1993) *Byggproduktmodeller – Nuläge*, Stockholm: Byggforskningsrådet (in Swedish).

Björk, B.-C. (1995) *Requirements and Information Structures for Building Product Data Models*, Espoo: VTT Offsetpiano.

Bunge, M. (1983) *Treatise on Basic Philosophy, Vol. 5: Exploring the World*, Reidel: Dordrecht, Holland.

Eastman, C. (1999) *Building Product Models: Computer Environments Supporting Design and Construction*. Boca Raton, FL: CRC Press LLC.

Fridqvist, S. (2000) *Property-Oriented Information Systems for Design*, Lund: KFS.

Gero, J. (1985) An overview of knowledge engineering and its relevance to CAAD, *Computer-Aided Architectural Design Futures*, Cambridge: University Press, p. 107-19.

O'Connell, D. (1983) *eCAADe, Digital Proceedings 1983–2000*, eCAADe (Education in Computer-aided Architectural Design in Europe), p. I.2.

Penttilä, H. (1989) *eCAADe, Digital Proceedings 1983–2000*. eCAADe. (Education in Computer-aided Architectural Design in Europe), p. 3.2.9.

Szalapaj, P. (2001) *CAD Principles for Architectural Design*, Oxford: Architectural Press.

Wikforss, Ö. (1993) *Informationsteknologi tvärs genom Byggsverige*, Solna: AB Svensk Byggtjänst (in Swedish).

Chapter 21

Using 4D CAD in the Design and Management of Vertical Extensions to Existing Buildings

Susan Bergsten

Introduction

The demand for apartments situated in city centres is presently high and is likely to increase. Extending existing buildings vertically and horizontally has been a frequently-adopted approach over the centuries in older cities. In some cities, apartments have been added to existing buildings such as shopping centres, offices and multi-storey car parks by vertical and/or horizontal extensions or conversions. However, poorly constructed vertical extensions have led to buildings collapsing and hesitancy by owners to contemplate such developments. Traditional methods of construction are generally used for vertically extending buildings, but are often not cost-effective (Andersson & Borgbrant 1998). Further development within larger and more compact cities must therefore make use of lighter building materials, novel building techniques and more efficient production processes. Even so, there is bound to be concern over the certainty with which newer materials, techniques and processes can provide an adequate and safe solution.

Designers are accustomed to producing mock-ups of the end product for communicating their ideas and, perhaps later, for production planning and control. Usually, these mock-ups are physical scale models, but increasingly 3D CAD systems are used to portray the end product. 3D cannot, however, take account of the production process without extensive adaptation. The 4D concept represents geometrical product (3D CAD) information together with process information (time) and offers a way forward for owners and designers who are considering complex additions to existing buildings.

This chapter describes the potential for the industrial production of light-gauge steel framing systems coupled with the use of 4D CAD. This is seen as a potentially cost-effective alternative for the vertical extension of existing buildings.

State-of-the-art review

Vertical extension of existing buildings

In most capital cities, and Stockholm is no exception, one of the most coveted places to reside is in the city centre. A consequence of this interest is the creation of more densely populated areas, where every possible space for accommodation is considered. In fact, many buildings in the historical parts of European cities have been extended once or even several times over long periods, as Fig. 21.1 shows. Bergenudd (1981) gives many examples of the vertical extension of buildings.

More recently, conversions and extensions to existing buildings have been successful in several countries; for instance the United Kingdom, Netherlands and Sweden (Verburg 2000; Hiller *et al.* 1998). Another example is that of the Robert L. Preger intelligent workplace, which has received an award for innovation under the auspices of IDEAS (Innovation Design and Excellence in Architecture with Steel Award 2001). This particular building is a one-storey extension of an existing university building in Pittsburgh, USA.

A survey by Bergsten & Wall (2002) on the vertical extension of existing buildings in Stockholm was undertaken during 2001. The results of the survey show that for vertically extended buildings, no special construction process or building methods were used. Many of these projects have been expensive

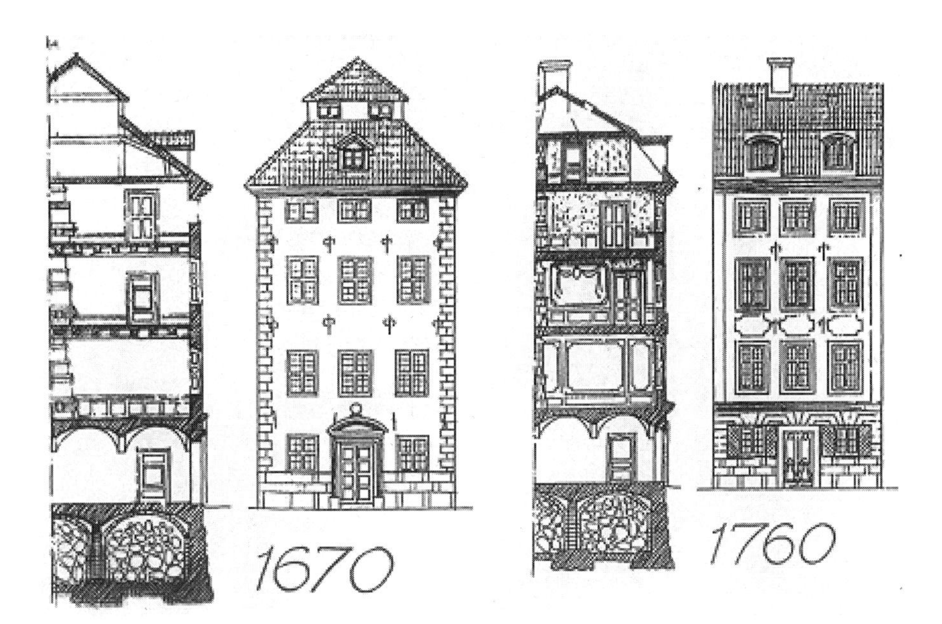

Fig. 21.1 Vertical extensions in the seventeenth and eighteenth centuries (Bergenudd 1981).

and only made possible because of a local boom in the construction sector. Three vertical extension projects have been studied in detail in this survey. The projects are: *Klarazenit* with steel column and cast in situ concrete slabs; *City-Cronan* with steel column and pre-cast concrete slabs; and *Berzeliipark* with a combination of concrete slabs, steel columns and light concrete and steel slabs. All these extensions offer between three and five additional storeys. In all cases, the existing building is of reinforced concrete. Studies of these projects have revealed many common problems as summarised below.

- The durability of the existing structure and foundation is a very important consideration during the design of the new structure. On-site tests are needed to confirm the quality of the construction. It is necessary to check carefully the local strength of attachment points between the existing building and the extension and, if necessary, strengthen them.
- Moisture control and weather-tightness (rainwater disposal and water-tightness) are essential requirements for reducing the risk of moisture problems during production and later in the use phase of the building. During the construction of vertical extensions some parts of the existing building will be exposed and therefore vulnerable to the weather.
- Working space allowances on the construction site must be planned before work commences in order to minimise problems from the lack of space due to the strict boundaries of the site.
- The construction plan must be communicated to the people affected by the work. It is not only for the benefit of those managing and working on the site: there are neighbourhood responsibilities too.
- Logistics planning from, to and on the site is very important. Also, the impact on traffic around the site, especially in city centres, has to be considered. At the same time the overall lack of space on the site leaves little space for storing materials.
- Properly planned logistics on, to and from the site are vital for efficient working.

As seen above, the problems related to vertical extensions are not uncommon problems in construction generally, but are perhaps more acute in these circumstances. Constraints on construction are often site- and geometry-related, such as in the vertical extension of existing buildings, where access and movement restriction and other physical constraints imposed by the existing building make the production process more complex. This implies that these constraints must be carefully considered during design decision-making.

In vertical extension projects, it is also important to minimise disruptions and eliminate hazards for neighbouring properties and legitimate activities especially in the vicinity of the site. When altering the existing environment,

Fig. 21.2 Modern vertical extension in central Stockholm.

several aspects must be considered and must be taken into account in the planning process:

- influences that the extension will have on the existing building;
- influences of the extension's activities on existing activities; and
- influences of production activities on the everyday activities of the existing building and its environs.

The trend in the 1960s and 1970s for flat roof construction has left its mark on the townscape. In many cities, there are large areas of flat roof, underneath which the existing structure has the ability to bear the extra loads from additional storeys (see Fig. 21.2). To make the best use of these areas, it is essential to develop building methods that are practicable, cost-effective and appropriate.

Light-gauge steel framing for vertical extensions

The main structural components of the light-gauge steel framing system are galvanised cold-formed steel sections. The system has been tested and is suitable for up to five storeys. Research and application have confirmed many benefits from the use of light-gauge steel framing in housing, some of which are mentioned below (Burstrand 2000; Gorgolewski *et al.* 2001; MacCarthy 1998):

- structural performance;
- ease of construction and deconstruction;
- lends itself to pre-fabrication;
- reuse and recycling of material;
- good level of sound and thermal insulation;

- dry construction process; and
- improves the chances of consistent quality.

Light-gauge steel framed buildings with their lower weight, when compared with other traditional buildings, have been recognised as suitable for vertical and horizontal extensions to existing buildings (Peterson & Öberg 2001). They outperform similar concrete buildings in terms of their weight by a factor of five (Burstrand 2000). Other experiences confirm the suitability of this method for vertically and horizontally extending existing building (Tomá 1999).

Another important attribute, as mentioned above, is its industrialised production method. The design and manufacture of light-gauge steel framing lends itself to pre-engineering for off-site assembly of elements, tight tolerances and simple site erection of the elements (MacCarthy 1998; Gorgolewski *et al.* 2001).

Information management and 4D CAD

The use of 3D modelling is an important aspect of the industrial production of light-gauge steel framing systems. In the past few years, the use of 3D modelling in design and the procurement of light-gauge steel framing systems has increased. The 3D model defines the product and shares a common database. This common database generates the latest version of drawings and information for use by different actors. Many kinds of documents can be generated from the model; for example, perspectives, material specifications, workshop drawings and assembling drawings. The database can also produce data for cost estimating, time scheduling, manufacturing and contract tenders (Cederfeldt 1997). The product and the production process must be considered during planning and scheduling. Linking a 3D model to the process is currently undertaken with the use of, for example, critical path method (CPM) schedules.

The existence of the 3D model and visualisation techniques are effective tools for representing production information in detail in order to shorten overall time and increase project productivity (Koo & Fischer 2000; Akbas 2001; Webb 2000; Leinonen & Kähkönen 2000).

The 4D concept can be described as a matter of connecting the 3D model to the production process and trying to visualise the production through the use of different colours – see Fig. 21.3 and, later, Fig. 21.4. By visualising and building the structure in the computer, prior to work on-site, the 4D model can help identify constructability and sequencing problems. Other benefits from 4D CAD are: better communication and co-ordination between project actors, conveying the spatial constraints of a project, foreseeing hazardous situations and safety matters. A 4D model also assists in visualising workflow on the

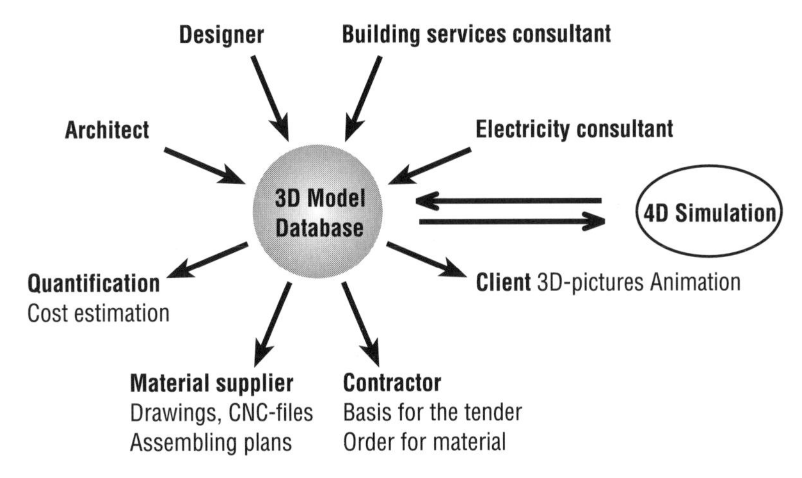

Fig. 21.3 4D CAD concept.

site and the allocation of resources and materials (Koo & Fischer 2000; Staub *et al.* 2000).

The results of a case study by Staub *et al.* (2000) show that today's mechanism for 4D model generation, adjusting the 3D model and linking it with the schedule, is too complicated for everyday use. Furthermore, many constraints are overlooked during detailed production planning. The time taken to create a model depends upon the application of the 4D method, the level of detail provided in the 3D model and the user's knowledge. A vital consideration for the project's design is how work is organised on the site, especially in the early phases. Also important is how work is brought on to the site and controlled. Attention to these matters during the design helps to avoid conflicts (Howell 1999). Currently, research is being conducted into the automatic generation of construction zones from 3D models to assist in construction planning and scheduling (Akbas 2001).

Despite the enormous benefits promised by 4D modelling, Koo & Fischer (2000) discuss some limitations in the method. Although 4D models can help relatively inexperienced users to identify problems in construction projects, they cannot convey all the information required for evaluating schedules and activities. Users can easily infer physical constraints from the 4D model, but non-physical constraints are harder to establish. Using different colours for showing different activities can mean that models become quite cluttered, leading to a loss of detail definition. Also, generation of rapid alternative scenarios is difficult and labour intensive. Besides, it is important to specify the types of operation and the level at which detailing a 4D concept provides the most benefit. Many contractors assume too quickly that the cost of CAD operators on top of 3D and 4D modelling systems is prohibitive. Research has shown that by using 4D CAD in design and construction, overall productivity in the project will rise and many other benefits will accrue to the actors involved (Staub *et al.* 2000).

Researchers who target the technical problems in 4D CAD often forget the impact and difficulties of softer parameters. Problems related to implementing *4D thinking* in the construction sector have been reviewed by Barrett (2000) and can be summarised as low organisational readiness, tacit-tacit emphasis, high action and reactive orientation and economic turbulence in the sector. In fact, these factors should be considered during the work of 4D CAD implementation. The efficient use of 4D CAD implies that the 4D concept and software are implemented correctly. Targeting the technical aspects and not the user aspects will result in disappointing experiences and a lack of realisation of the primary benefits of 4D modelling.

4D CAD practice in construction sector

Today, there are many examples where virtual reality and simulation techniques have been used successfully as, for instance, in the shipping industry. The product model is used throughout the design, fabrication and assembly phases, and 3D/4D simulations with connections to numerous databases (including those for cost and time) have shown themselves to be a reliable way of increasing productivity and competitiveness (Douglas 1994). Other examples in the heavy engineering and process industries can be found.

The most frequent use of visualisation and 4D CAD in construction has been in the marketing and pre-construction phase. Some attempt has, however, been made to facilitate visualisation during the construction process. An early attempt by Bengtsson & Bergstrand (1999) to apply the 4D method in the construction phase was by manually connecting different layers in a 3D model to a time schedule. Different time sequences in the production process were introduced to different layers. By revealing or suppressing layers, various sequences of the production could be visualised (see Fig. 21.4).

More sophisticated attempts to implement the 4D concept in the construction phase have been undertaken by the Center For Integrated Facility Engineering (CIFE) at Stanford University in the US, and have attracted industrial interest.

Today, there are several 3D and 4D modelling tools on the market that have been used by contractors (e.g. Bovis using Bentley) in pilot projects or case studies (Webb 2000; Leinonen & Kähkönen 2000). In Norway, a 4D model has been used for the project *Pilestaedet Park*, using a 3D model and time planning program. Other attempts at realising 4D methods have been seen in the integration of life cycle data with building and visualisation (Linnert *et al.* 2000).

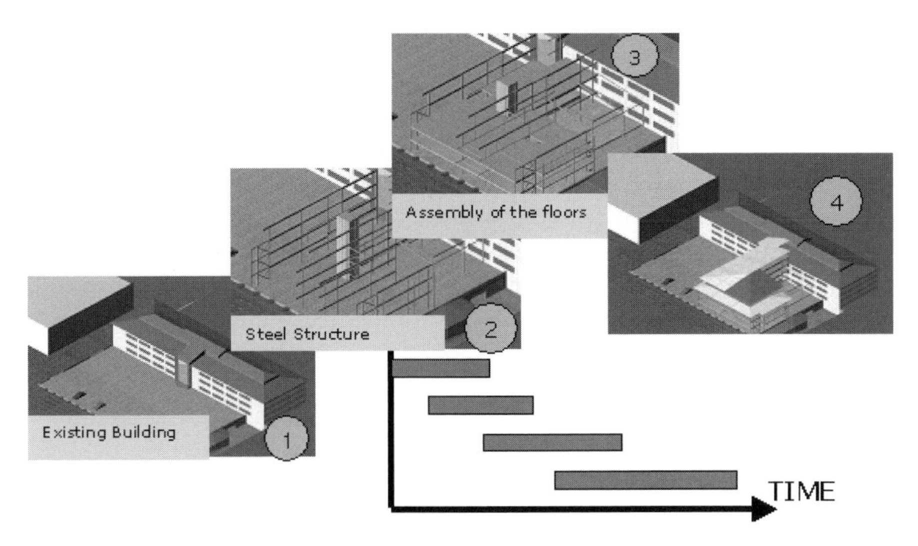

Fig. 21.4 Example application of 4D in construction.

Using 4D CAD for vertical extensions

Research indicates that the light-gauge steel framing system is a cost-efficient alternative for multi-storey housing projects. Using 3D modelling systems for design and production planning in the industrialised production of light-gauge steel framing systems increases productivity on the project (Burstrand 1998; Cederfeldt 1997). Case studies reveal that light-gauge steel framing, together with 3D modelling, have reduced production costs by approximately 20% for multi-storey housing projects and increased overall project productivity (Andersson & Borgbrant 1998). Usually, contractors want to accelerate the on-site works. In order to minimise construction time, industrial production methods can be used (Fernström & Kämpe 1998), although accurate planning of resources, space and activities should be done early in the project and not, as often is the case, for overcoming delays. By using 4D models and increasing the reliability of the project schedule, an efficient route to more productive and efficient construction process can be found (Akinici *et al.* 1998).

Combining the use of light-gauge steel framing with 4D modelling also succeeds in bringing a degree of co-ordination and integration to the design and production process that might otherwise be absent. 4D methods enable the design team to simulate the co-ordination of the extension with the existing structure. By considering constructability, production methods, interdependency of tasks and matching manpower to available work in the design phase, changes and inaccuracy can be minimised. This will also minimise some of the disturbance caused to the surrounding area.

Visualisation also assists the project team by showing the status of the project at any time. For example, the project can be simulated for the benefit of neighbours of the site in order to inform them about the progress of the project and how the project will affect them during different periods. Alternative designs for the extension can be simulated in order to determine the most process-efficient solution. Logistical considerations, such as access to the site for delivery vehicles and materials handling on the site, can be introduced, and different scenarios tested. Lastly, the impact of on-site works on traffic flows can be investigated in order to minimise disturbance in the vicinity of the site.

Research project

Project description and objectives

Extending buildings vertically, especially in city centres, is fraught with technical and managerial problems. Many of the problems have been mentioned earlier and are particularly worrisome with respect to apartment buildings. Since apartments, as compared with commercial buildings, have the highest requirements for sound insulation and fire protection, they provide a demanding test-bed. The results of the research could also be applicable to other building systems used for housing and commercial and industrial buildings. Against this background, the aim of the research is to identify cost-effective production methods for the vertical extension of existing buildings. The industrial production method of light-gauge steel framed systems, together with the use of 4D CAD, is being investigated in order to support the achievement of this objective.

Another aim is to produce guidelines for integrated design and production methods when industrialised building methods are used. These will be based on the most appropriate means for simulating industrialised production methods and the implications of a given design.

Josephson & Hammarlund (1999) found that the majority of the defects in construction could be ascribed to design, site management, subcontractors, materials and execution of the work. This research project will therefore determine if the 4D concept can help in minimising defects and thereby increasing a project's overall productivity. The productivity of the project will be measured using the method discussed by Jansson (1996).

Research methodology

Case studies will be used as the primary approach for investigating new methods and new design and management tools (Yin 1994). A study of the design and production planning process within the respective projects will be carried out in close co-operation with the owners, designers and contractors. One case study is being undertaken with respect to a highly industrialised construction process using light-gauge steel for the vertical extension of an existing building in Stockholm. This study will also evaluate the design and the planning process and the physical result with respect to quality, cost and customer satisfaction. The usefulness of applying 4D modelling will be determined from this work and can be summarised as:

- evaluating 4D modelling in comparison with the actual procedures for design and planning; and
- drafting guidelines for implementing *4D thinking* in the Swedish construction sector.

Research results and industrial impact

Tentative results

The expectation is that the results of the case studies will largely confirm the utility of applying 4D modelling, especially to the problem of the vertical extension of existing buildings. This will be based upon industrial production methods using light-gauge steel framing. It is further expected that the case studies will provide hard evidence of the extent to which this approach is able to minimise the duration of work done on the site and disturbance to the neighbourhood. Evidence of a lower than normal requirement for storage space on-site will also be expected to emerge.

Implementation and exploitation

The application of the approach outlined in this chapter will also be considered for application to other types of building. Moreover, it is expected that the results of the research will spin off into other areas of construction activity. There appears no reason why 4D modelling could not be deployed on other kinds of building and construction problem. Other uses for light-gauge steel framing systems are likely to receive some measure of support from the successful completion of this project.

Conclusions

The problem of extending existing buildings vertically, especially to provide space for apartments, has been discussed in this chapter. Two strands of research are being pursued and attention has been drawn in the earlier sections to the current state-of-the-art in the application of 4D CAD modelling. This has been set against the background of the availability of light-gauge steel framing systems that potentially offer an attractive and speedy solution for added accommodation in city centres and other densely populated areas. The evidence so far collected would seem to confirm that the industrial production of the light-gauge system, together with the use of 4D CAD modelling to support production planning, would provide greater certainty of success for projects involving vertical extensions. The case study method adopted in the research is expected to provide specific evidence of the practical use of the approach as well as highlighting changes that are needed within the construction sector for it to gain acceptance.

References

Akbas, R. (2001) Formalizing domain knowledge for construction zone generation, *Proceedings of the CIB-W78 International Conference IT in Construction in Africa*, Mpumalanga, South Africa, pp.30-1–30-16.

Akinici, B., Fischer, M. & Zabelle, T. (1998) Proactive approach for reducing non-value adding activities due to time-space conflicts, *Proceedings of Sixth Annual Conference of the International Group for Lean Construction (IGLC-6)*, Guarujá, São Paulo, Brazil, pp.1–16.

Andersson, N. & Borgbrant, J. (1998) *Hyreskostnad, förvaltning och producktion i harmoni*, Luleå: Luleå University of Technology, Department of Construction Management, Research Report 1998:12 (in Swedish).

Barrett, P. (2000) Construction management pull for 4D CAD, *Proceedings of the Construction Congress VI Building together for better construction in an increasingly complex world* (ed. Walsh, K.D.), ASCE, Orlando: Florida, pp.977–983.

Bengtsson, B. & Bergstrand, J. (1999) Byggprojketering i 4D, *Väg och Vatten Byggaren, 5*, pp.12–15 (in Swedish).

Bergenudd, C. (1981) *Påbyggnader*, Doctoral thesis, Lund: Lund Institute of Technology (in Swedish).

Bergsten, S. & Wall, M. (2002) *Påbyggnader i början av nya seklet*, Stockholm: Swedish Institute of Steel Construction, SBI Technical Report 227:1 (in Swedish).

Burstrand, H. (1998) Light-gauge steel framing leads the way to an increased productivity for residential housing, *Journal of Constructional Steel Research*, Paper No.213, 46:1–3.

Burstrand, H. (2000) *Light-Gauge Steel Framing for Housing*, International Iron and Steel Institute (IISI), Västervik: AB CO Ekblad & Co.

Cederfeldt, L. (1997) *Modellorienterad Projektering av Bostadshus med Lättbyggnad – Kv. Näktergalen, Ängelholm*, Stockholm: Swedish Institute of Steel in Constriction (SBI) (in Swedish).

Douglas, B. (1994) A ship product model as an integrator between vessel building plan and design, *Proceedings of the 8th International Conference on Computer Applications in*

Shipping, Vol. 1. (ed. Brodda, J. & Johansson, K.), Bremen, Germany: Bremer Vulkan Verbund AG, pp. 35–45.

Fernström, G. & Kämpe, P. (1998) *Industriellt byggnade växer och tar marknad*, Stockholm: Byggförlaget (in Swedish).

Gorgolewski, M., Grubb, P.J. & Lawson, R.M. (2001) *Modular construction using light steel framing: design of residential buildings*, SCI-P-272, Ascot: Steel Construction Institute.

Hiller, M., Lawson, R.M. & Gorgolewski, M. (1998) *Over-roofing of existing buildings using light steel*. SCI-P-246 Ascot: Steel Construction Institute.

Howell, G.A. (1999) What is Lean Construction? *Proceedings of the 7th Annual Conference of International Group for Lean Construction (IGLC-7)*, Berkeley, California, USA, pp. 1–10.

Jansson, J. (1996) *Construction Site Productivity Measurements, Selection, Application and Evaluation of Methods and Measures*, Doctoral thesis, Luleå: Luleå University of Technology.

Josephson, P.-E. & Hammarlund, Y. (1999) The Cost of Defects in Construction. *Automation in Construction*, **8**, pp. 681–7.

Koo, B. & Fischer, M. (2000) Feasibility Study of 4D CAD in Commercial Construction, *Construction Engineering and Management*, **126**, 4, pp. 251–60.

Leinonen, J. & Kähkönen, K. (2000) New construction management practice based on the virtual reality technology, *Proceedings of the Construction Congress VI Building together for better construction in an increasingly complex world* (ed. Walsh, K.D.), ASCE, Orlando: Florida, pp. 1014–22.

Linnert, C., Encarnacao, M., Strok, A. & Koch, V. (2000) Virtual Building Life-cycle – Giving Architects Access to the Future of Buildings by Visualizing Life-Cycle Data, *Proceedings of the Eighth International Conference on Computing in Civil and Building Engineering*, Vol. 1, Fruchter, R. (ed.), Reston, VA: ASCE, 14–16 August, pp. 7–14.

MacCarthy, I. (1998) Prefabricated building method using cold-formed steel components, *Journal of Constructional Steel Research*, Paper No.417, 46:1–3.

Peterson, M. & Öberg, M. (2001) *Multi-dwelling concrete buildings in Sweden*, Report TVBM 7158, Lund: Lund Institute of Technology.

Staub, S., Fischer, M. & Spradlin, M. (2000) *Industrial Case Study of Electronic Design, Cost and Schedule Information*, Stanford: Stanford University, CA, Center For Integrated Facility Engineering (CIFE), Technical Report No.122.

Tomá, A.W. (1999) *The application of steel in urban habitat: extra storey on existing building*, Delft: Netherlands Organisation for Applied Scientific Research (TNO), Report CON-BIS/R5019/1.

Verburg, W.H. (2000) *Bouwen op toplocaties*, Rotterdam: Bouwen met Staal (in Dutch).

Webb, R.M. (2000) 4D CAD-Construction Industry Practice, *Proceedings of the Construction Congress VI Building together for better construction in an increasingly complex world* (ed. Walsh, K.D.), ASCE, Orlando: Florida, pp. 1042–50.

Yin, R.K. (1994) *Case study research – design and methods*, 2nd edition, Thousand Oaks: Sage Publications.

Chapter 22

Importance of Architectural Attributes in Facilities Management

Ulf Nordwall

Introduction

There is growing awareness of the importance of architectural quality to facilities management and of the importance of non-measurable, aesthetic attributes to the home and its occupants. For too long, practical and functional aspects have dominated discussion of architectural quality in many fields including, albeit more recently, facilities management. The results of this can be seen in our homes. There are many apartments and houses in which the occupants, despite measurable qualities, find it hard to feel satisfaction and make a home. Slightly less than 40% of Sweden's population live in multi-family apartments (Andersson 1997) and it is this category of occupant that represents the highest number of people who are dissatisfied with their home.

There are different reasons for this dissatisfaction, such as poor maintenance, poor sound insulation, insufficient service and insecurity. The aesthetics of the neighbourhood and the building also have a prominent position in the evaluations that have been made. Statistically, those who live in apartment buildings – often and rather crudely termed blocks of flats – are four times more dissatisfied with the aesthetic qualities of their home in comparison with those who live in single-family houses. For the housing companies, this type of dissatisfaction can result in vacancies and, consequently, financial problems.

Many of the measurable, practical, architectural attributes in facilities management – the functional attributes – cover what we can physically delimit, measure and quantify. Practical attributes are conscientiously described in housing research, that has been carried out in Sweden since the 1930s. This information has been compiled into volumes of standards that are used in designing today's housing. These standards provide information on the home's practical functions; in other words, accessibility, fittings, physical attributes such as heating and ventilation, as well as the design of the external environment. The non-measurable architectural attributes in facilities management are the qualitative aesthetic and symbolic attributes, that are vital to the individual's perception of the home and its place in an apartment building.

An implicit aim of facilities management is to maintain and modernise a building so as to extend its usefulness and/or lifetime. Such an aim entails developing knowledge and questions about the conditions that promote a building's architectural attributes and which conditions counteract them. A wide range of conditions applies here: architectural, ideological, financial, technical or conditions affecting the use of the building. If one such condition points to demolition, it may suffice to demolish the building. The building can perhaps be seen as unfashionable or maybe it symbolises something negative, even though it can be used and is technically and financially sound. During the progress of the research, another significant condition has been added, namely a building's changing function over time.

The purpose of this chapter is to examine the connection between architecture and facilities management. Issues discussed include the significance of physical attributes and the architecture of a building, when it is to be managed over a long period. Several research questions are addressed; for example, which attributes of an apartment building are important to the facilities manager and the user respectively? Do architects and occupants desire the same architectural attributes? This study has two perspectives: one is to examine the connection between architecture and facilities management, and the other is to reveal architectural attributes that are significant from a facilities management perspective. The objective of the research is to provide more knowledge about these and other architectural attributes and the role they play in facilities management. The findings of the research could contribute directly to revealing the architectural attributes that are significant from a facilities management perspective, as an important step in adding value for the customer. This implies their proper consideration during the design phase.

State-of-the-art review

The research, which covers the evaluation of architectural attributes and how they impact on the process, is wide-ranging and involves many different disciplines. Literature studies, that underpin the research, have been undertaken in both the architectural science and social science fields.

Managing housing

Many different actors are involved in a building's planning and realisation before it enters into the occupancy phase of facilities management. A common metaphor for this combined building and management process is the relay race. The process entails specifications of the building's space and functions during the concept phase; while design, construction and choice of materials take place in a later, scheme design phase. By the time the building has been realised according to its specification and has been turned over to the facilities manager, many actors have been involved with each passing the *baton* to the next in line. The communication of information and the interpretation of drawings are notoriously problematic. Lawson (1991) questions if the original vision, with which the architect had to work, really survives this process. Lawson points out the importance of all actors in a building project being familiar with each other's solutions and having a clear understanding of the final product. It is easier to understand and deal with an event that is somewhat limited in time, than a long process in which control shifts from one actor to another. It is, however, during the occupancy phase that a building's functions, durability and other architectural attributes are ultimately put to the test, with time as the judge of how the building will fare.

The annual cost of operating, maintaining and repairing housing in Sweden – a country with a population under 9 million – is estimated to be SEK 66 billion (€7.3 billion) annually (Statistics Sweden 2001). By anyone's standards this is an enormous figure, and one that is likely to be many times higher in larger economies. Finding ways to reduce these costs could lead to substantial social profits for society. Furthermore, of a total national wealth of approximately SEK 4721 billion (€523 billion) in 1995 (Statistics Sweden 2001), Sweden's building stock accounts for approximately SEK 1493 billion (€165 billion). Since building components wear out and have to be replaced, most surfaces of buildings will have been replaced or changed in some way after 20 years. Replacement of this order has an impact on total national wealth (Antonsson & Lundin 1981).

Buildings are planned, designed and constructed over relatively short periods, yet they are required to be used for a long period thereafter. Effort is normally concentrated on the short-term creation of the building or the equally short-term, temporary reconstruction in which room layouts are changed or other elements, for example stairwells, are rebuilt.

Energy consumption, cleaning, the use of chemicals and waste disposal, together with transportation that consumes fuel and emits greenhouse gases, also take place during the facilities management phase. This implies that there is arguably greater potential to save money during the facilities management phase than in the planning, design and construction phases (Björkholm & Svane 1998). The facilities management phase is between 25 and 100 years. Seen from another perspective, this is a matter of running one of the country's most important sources of social capital.

Aspects of culture are carried into the future through the buildings and urban landscape that surrounds us. We influence our buildings by using them, by modifying them and through repairs and daily wear and tear (Werne 1987).

Facilities managers are nowadays responsible for the changes that take place in our buildings and sometimes we can see that our building environment can be rebuilt many times before it finally breaks down (Lang 2001).

Facilities management, including maintenance, is a process that is constantly being performed. The process is so common and routine that it can pass unnoticed for those not directly involved – almost an invisible service. This process is noticed more when it is not undertaken properly than when it is done correctly (Lundgren 2001). Correct maintenance is necessary for a building to function satisfactorily for users, as well as satisfying facilities managers. Insufficient maintenance results in the building decreasing in value and in future measures being more complicated and almost always more costly (Lundgren 2001). Poorly undertaken maintenance can influence the self-esteem and comfort of those who live and work in the building. Taking care of buildings also has to do with caring for people's living environment. Despite this, maintenance is often considered to be a secondary work process (Rönn 1989).

During the early to mid-1990s, when construction was at a low level in Sweden (and many other countries), the focus shifted from new construction to an increasing interest in the existing building stock. Good facilities management has become the basis for competition and, by extension, this means that it is easier to rent out or lease well-managed apartments. Occupants also tend to take better care of such buildings and normally stay for many years (Lundgren 2001). Examples of clear, distinct changes in the management of apartment buildings are found in the following statements by Junestrand (1998), stressing the importance of:

- a distinct focus on the occupant as customer and an emphasis on long-term customer relations;
- continuous rationalisation and cost monitoring; and
- better and modified skill requirements for personnel.

And, by Lundgren (2001):

- being aware of the significance of IT in the development of facilities management; for instance, computerising monitoring systems;
- improving communication between occupants and the facilities manager, through the provision of quicker information;
- initiating more self-management;
- cultivating the housing stock; and
- investing to reduce operating costs.

From a longer-term perspective, it is possible to imagine many different organisational models and methods of working for a facilities manager (Junestrand

1998). The greatest interest for a facilities manager is to develop a long-term relationship with occupants/users (Lundgren 2001).

Political and social dimensions

Ethnologist Karl Olov Arnstberg has studied and discussed, over a long period, housing development from the Swedish building boom in the 1960s, commonly called the Million Programme. In his book, Arnstberg (2000) summarises texts and ideas he has been working with in recent years. The result is a piercing criticism of Swedish housing policy as well as of the corpus of Swedish architects. He describes the Million Programme as a gigantic building experiment in which the structure was important and not the human being. Arnstberg also presents facilities management as a new area of activity for architects. Once progressive housing contractors have realised the value of successful facilities management, architects must also realise the value of this activity. What do ageing, history, durability, occupant awareness and self-management mean to the architect, Arnstberg asks?

Housing was constructed rationally during the Million Programme. Despite this, some of the housing has become very costly over the longer term. Arnstberg complains that the construction sector is still much more conservative than the facilities management sector. Construction companies still have the attitude that it is the building that is important, not the human being, and companies build with technology instead of building for people. Many of the housing areas from the Million Programme have been rebuilt once, twice and even three times, and still the problem of vacant, unattractive housing has not been solved. This is an example of poor economic practice for which occupants must pay (Solberg *et al.* 2001).

In the Housing Bill of 1998, the Swedish government stated that the country was essentially completely built after decades of large-volume new construction. The Bill maintained that production policy was becoming more and more like housing policy and that priority was to be given to existing housing areas. Considering the total lifetime of a building, the facilities management phase represents up to 70–80% of the total cost (Fall 1999). Investments made during the facilities management phase are many times higher than the acquisition value (Jensfelt 1994). The greater focus on facilities management also depends on managers' increased awareness of competition if they want to make their housing more attractive in order to retain occupants (Lundgren 2001). Knowledge of architectural attributes is therefore important from a management perspective in order to create better conditions for qualitative, visionary and profitable construction. The maintenance of buildings takes place on many different levels, according to Lundgren (2001):

- Operation is about running a building from day-to-day. This may be adjusting a time lock, changing a broken light, taking care of rubbish or cleaning the stairwell.
- Repair or emergency maintenance is about fixing sudden defects and problems, such as a leaky tap or a blocked drain.
- Planned maintenance entails slightly larger measures that are performed at regular intervals to maintain a certain standard for the building. An example of this is sanding a parquet floor. It is possible to predict the measures that need to be applied. However, it may be difficult to predict exactly when they will need doing, since the intervals for certain measures are very different from building to building depending on location, original quality and use.
- Extraordinary maintenance calls for measures that deviate from expected maintenance. This may depend on damage or the discovery of so-called hidden defects. The border between measures is blurred. Normally, however, decisions on measures are made on different levels. Repairs are undertaken quickly by the building's custodian and repairmen.
- Periodic maintenance is planned with the person responsible for facilities management.

Many years may pass before a particular building is in need of attention. Even so, a major housing company is likely to perform a multitude of measures each year, across its housing stock. None of the measures raises the question of quality *per se*, as they are largely concerned with maintaining the existing condition. Costs are covered by current rents (Benjamin 1996).

Reconstruction, in contrast to maintenance, is not continuous. It is extensive and aims at raising quality, normally to a level corresponding to a new construction standard. This entails major investment, that often leads to a change in the occupants of entire buildings. Changeover of occupants can have a major social impact, which is not always desired by the occupants or the facilities manager (Hurtig 1995).

30-year reconstruction has long been an established concept. The assumed lifetime of services installations have generally pre-determined intervals. After 30 years, the main distribution lines and pipework are considered worn out and the ventilation system usually requires renewal. Improvements are often simultaneous with the replacement of other elements, for example roofing felt (Fall 1999). Measures tend to be major and require building permits and associated specifications (Lundgren 2001). Technical requirements for various attributes must be satisfied, including fire safety and accessibility for people with impaired movement and orientation ability. The facility manager or owner, at this time, often takes the opportunity to renew and change other parts and systems (Lundgren 2001).

Occupants are offered alternative housing during the reconstruction period. With reconstruction, the owner renegotiates the rent with the tenants' association on behalf of all occupants, something that is not done for minor measures such as maintenance. The Utility Value Act assumes a standard price for a certain number of rooms plus kitchen, a price that is then modified upwards or downwards, depending upon the building's location in the town and the apartment's location in the building – if there is a lift, a balcony, a washing machine in each apartment and so forth. It is possible to argue for architectural qualities, but this is seldom accomplished, and it is generally easier therefore to argue for technical equipment.

The national financing system, including the renovation and extension grants normally designated as *ROT*, that existed in the 1980s, often encouraged housing owners to collect a number of minor measures into one major measure. For a facilities manager, it could be cheaper to tear out kitchen fixtures than to renovate them. It was appealing to take advantage of this opportunity, and the new fixtures were supposed to last longer than those being removed (Lundgren 2001). Ironically, the results were not always what was expected when untested material was installed. Cupboard doors made of melamine, for example, turned out to be impossible to repair and were quickly exchanged (Niklasson 2001).

Vidén (1990) reveals that reconstruction undertaken mostly during the 1980s did, in many cases, exceed the building's purely technical needs. What is worse is that occupants, especially those in housing areas dating from the 1940s and 1950s, often felt that the measures had been unnecessarily extensive, that many appreciated qualities had been removed and that they had been given little opportunity to voice their opinions in general and had, therefore, been *run over* (Hurtig 1995).

According to thinking in the USA, where the concept of facilities management has been used since the 1980s, management is taken to mean a type of service company that sells the service of *living* to users who are the customers. Williams (1996) maintains that of all management disciplines, facilities management is the one with the most potential for further development. There are, however, risks that conditions for good management can be poor, if ambitions to save time and money are too high and are not compensated for by seeking new work methods, such as continuous training or finding new norm systems to follow (Sweeney 1996). When occupants become the focus of attention, the building can no longer be regarded as simply a technical product to keep out the rain and cold; it becomes a home, that must take on the form the occupants desire (Fall 1999).

Apartments can differ significantly from one another. A basic standard can always be offered, but different options provide freedom of choice and a feeling of homeliness. People seek different types of homes during different stages of life. This must affect facilities management, that should be under continuous development (Jensfelt 1994). If the bicycle storeroom is located at an inconvenient distance, there is a risk that occupants will take their bikes through the stairwells and apartments and out on to the balcony, an action that soils, damages and wears out surface material.

A building constructed of material that ages well, allows greater freedom in terms of maintenance. An old brick wall is beautiful, while a peeling sheet metal facing must be repaired immediately (Jensfelt 1994, 1996). If a building falls into disrepair for a longer period, extensive repairs will most likely be required to restore the building to a condition suitable for habitation again. Such measures will be major and perhaps even technically complicated as well as expensive. The risk that the original architectural values become corrupted rises. Continuous maintenance extends the lifetime of different parts of the building, but gradually damage will arise which must be repaired. These repairs can be done in many different ways, depending upon the level of concern over the building's character and cultural or historical value. The problem that the facilities manager sees in the short term is that continuous maintenance is costly, but to avoid undesirable and drastic situations later, it is necessary to inspect buildings on a regular basis to ensure that they are continuously maintained. Measures must be taken before major damage occurs (Niklasson 2001).

Occupants can have an impact on facilities management; for example, through their association. Even the landlord can increase the users' potential to influence management, both collectively and individually. Another possibility the users have to influence facilities management is through the tenants' association, which meets a few times each year. When a building is to be reconstructed or extended a meeting is always held (Lundgren 2001).

Research project

Project description and objectives

The main purpose of this research is to identify the relationship between architecture and management and to describe a number of architectural attributes that are influenced by the selection of facilities management strategies. The identification of these attributes is needed in order to show how they impact on a building's lifetime and ageing, thereby providing more knowledge and understanding of important aspects of a building's long-term value. The central questions in the research are:

- Which architectural attributes are significant to the management of the facilities?
- How can management of the facilities preserve or develop these architectural attributes?

The preliminary sets of architectural attributes are: durability, change, renewal, material, execution and planning.

Research methodology

The research involves collecting facts and knowledge about the relationship between the three variables: architecture, management and occupants. As Jensfelt (1996) argues, these are considered best obtained through observations and practical studies. In order to discern the relationships between these variables, case studies have been used. While case studies have thus been selected as the primary method of enquiry, central to this approach is that hypotheses are generated and created from data that have been systematically collected from fieldwork. Data consisting of findings from interviews with occupants, architects, managers, contractors and municipal authorities etc. have been used to formulate and test the hypotheses. The application of Grounded Theory Method is significant in this respect, since the research is not based on preconceived ideas or theories (Barney & Strauss 1967). The specification of any hypothesis is a final product rather than a pre-condition for the research (Starrin & Svensson 1996).

Triangulation is a concept that is particularly helpful to researchers, not least in helping to confirm or substantiate findings, and has been adopted here. Different methods, such as interviews, observations and physical examples, are combined to support examination of one and the same unit (Merriam 1994). The method generates data, that contains as many dimensions and qualities as possible in order to test and develop theories (see Fig. 22.1).

The qualitative approach seeks concepts and quality contexts, based on observations. Here, it has the purpose of revealing architectural attributes that are important from the facilities management perspective. Twenty-five interviews were carried out during 2000 and 2001 among architects, contractors, facilities managers and occupants. The resulting theory describes aspects of

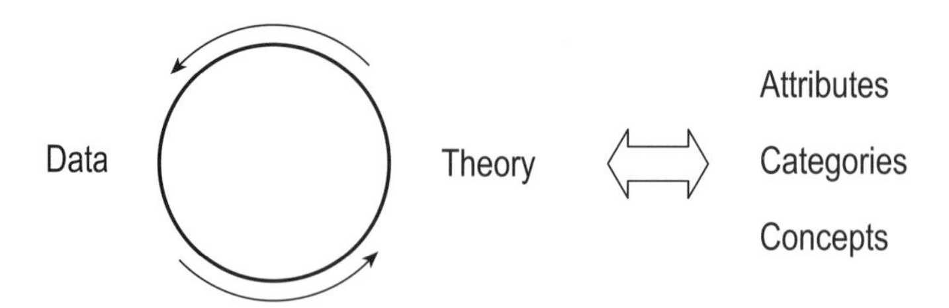

Fig. 22.1 Creation of categories, attributes and concepts through the interaction of data collection and theory.

those architectural attributes that are important from a facilities management perspective.

Research results

The study was based on six sets of attributes: durability, change, renewal, material, execution and planning. These different fields of attributes can be said to represent different scientific disciplines in the traditional sense: the humanities, business management, engineering and architecture. The purpose of each set of attributes is to formulate relevant and treatable problems, as well as to discuss, compare and develop methodologies. In this respect, the attributes raise as many new concerns as they address existing questions.

Durability

Various choices (material, technology, maintenance, reconstruction/renewal etc.), that determine a building's durability, are often based on financial motives. Durability, in the technical sense can, somewhat simplified, be described as durable in time (i.e. desirable technical attributes are maintained over time). It is possible to talk about the durability of a particular material, of structures (load-bearing and non-load-bearing), building components or outer surfaces. Durability often determines a building's way of ageing and, consequently, the architecture and its expression. Values of durability are significant to the way we see buildings and their lifetimes. Today, the management of buildings has gained greater attention and apartment buildings are seen as an asset to a greater extent than before. The values that steer the management of existing building stock are also vital to durability. Not only must management, with its eye to maintenance and change in use, be constantly placed in relation to the value of apartments, but also social, cultural, historical, community and symbolic values are essential to the lifetime of the existing building stock.

Values pertaining to durability are closely related to the same basis of valuation as facilities management issues. This mutual basis of valuation is being studied to be able to distinguish *what* is mutual as well as that which is specific. The change in values that takes place over time is central to the interpretation and understanding of why different buildings can be perceived as outdated, despite having qualities such as good technical and material durability (Rybczynski 1988). Architectural styles and expression change with time and our perception of style and expression changes continuously. A building's material durability can, often, give way to aesthetic changes in a valuation system. Social values of what our homes should be like and how we should live can mean that a building's durability is re-evaluated. What time perspectives are expressed in our view of durability and resource conservation during different eras? How has our view of durability shifted as it pertains to material and building engineering?

Change

A building is erected for a special purpose and is normally expected to last a long time. The original use will sooner or later change (Jensfelt 1978). Even if the building's function remains, the building will be outdated when modern ways of living demand new rooms, fittings and services installations. Sometimes, the building's original function becomes entirely obsolete. It can then be changed and used for new activities and, if this is not possible, it will either fall into disrepair or be demolished.

The potential and the desire to modernise, reconstruct and reuse a building determine its service life. For this reason, a building must have a certain degree of universality, flexibility and potential for reconstruction to survive changes in use (Berg 1994). Methods and principles for how extensions and reconstruction can be executed should be continuously developed; for example, ideas about how old meets new in architecture (Cold 1989). This matter has both aesthetic and functional aspects that carry different weights depending on the type of project.

Research and (artistic) development follow two lines: one that is based on the building and one that is based on the activity housed in the building. Can we distinguish attributes in new production that are vital to future changes in use? What ideas about the relationship between old and new in architecture lead to good use of resources and good management? How can we design a planning process that tests a building's attributes and the specifications of an activity, to find an optimum solution from the perspective of facilities management, architecturally and in terms of activity?

Renewal

The method used to solve an architectural problem, for example, executing details, affects the management of the building. What attributes of a building, its parts and building materials, are worth striving for from the perspective of facilities management and sustainability? Is it possible to exchange parts of a building and replace them gradually as needed? Do we strive for strength and resistance or yield, the ability to escape movement caused by land settlement, heat and damp? How do we achieve these attributes? Tough requirements for a perfect outer layer and right angles mean that we often repair before it is technically motivated. In what way do modern demands for smoothness and perfection affect technical solutions? Buildings can be repaired by unsuitable methods due to a lack of knowledge of the technology and material from which the building was once erected. What knowledge about old and new building engineering is relevant today, and how do we access this knowledge?

Materials

In earlier times, building materials were expensive in relation to the cost of labour. Today, we generally have the opposite relationship. Producing materials and buildings is labour-intensive.

In the past, limestone was carefully selected before firing. Timber was used rationally: the core was for vulnerable parts of a building, such as door and window frames and casements, and the outer layer was used for panels. Demolition material was reused. Pre-industrial and early industrialised building engineering was different from today's building.

The shift from craft-based work to more industrialised processes has had an impact on materials and construction, in terms of off-site manufacture and work performed on the construction site. In building there has been a discernible move away from yielding materials and technology, to unyielding – firmness, hardness, strength and insulation have gained greater importance. Materials traditionally used in construction, i.e. stone, wood, iron, brick and lime have, in the twentieth century, been used in combination with many new materials. Different kinds of plaster, boards and insulation have replaced and supplemented traditional materials. Some of these products are an imitation of genuine, traditional materials, yet they do not always perform as well as those they imitate (Werne 1993).

Execution

Management is also affected by how building materials are worked, how they are joined, by the degree of care used in execution and by accessibility for maintenance. Different parts of a building have different lifetimes and must be repaired or renewed within different time periods. What significance does the choice of material and details have from a facilities management perspective? The maintenance of a building must be based on an understanding and knowledge of the existing building, the material used and the various craft and production methods adopted. This knowledge is vital to the management of existing building stock, as well as to new production. In what way can execution and detailing influence and provide for better and more sustainable construction and facilities management?

Planning

The organisation of space in an apartment building means the organisation of the rooms in the building and the relationship between them. In what way(s) can the relationship between the home and utility spaces in the home influence patterns of movement, wear and tear, facilities management etc?

Conclusions

Thus far, the results of the research have led to the identification of six sets of attributes with non-measurable qualities. These attributes are durability, change, renewal, material, execution and planning, and represent the tools with which the research questions are being progressively analysed. Non-measurable architectural attributes are providing insight into and knowledge of the relationship between architects, occupants and housing companies in terms of the design of apartment buildings, and the occupants' relationship with their home. The central tentative conclusion of the research is that non-measurable architectural attributes in facilities management are significant to an occupant's perception of a home, and they are intimately connected to the occupant's process of appropriation and creation of meaning. Another conclusion is that the facilities management aspect of architectural attributes can, in the future, be a basis for competition and can give the building a value in the same instant as it is completed.

References

Andersson, Å.E. (ed.) (1997) *Bostadsmarknaden på 2000-talet*, Stockholm: SABO (in Swedish).

Antonsson, U. & Lundin, S. (1981) *Svensk Arkitektur 1900–1930*, Gothenburg: Chalmers University of Technology (in Swedish).

Arnstberg, K.O. (2000) *Miljonprogrammet*, Stockholm: Carlsson (in Swedish).

Barney, G. & Strauss, A. (1967) *Grounded Theory Method*, New York: Aldine.

Benjamin, D.N. (1996) *The Home*, Aldershot: Avebury.

Berg, K. (1994) De historiska dominanterna: Staden som bomässans förutsättning. In: *Framtidens vägvisare*, Bo 93, Karlskrona: Swedish Board of Building, Planning and Housing, pp. 13–14 (in Swedish).

Björkholm, Y. & Svane, Ö. (1998) *Miljöarbete i bostadsförvaltnig: från mirakelresa till miljöledning*, BFR T7:1998, Stockholm: Swedish Council for Building Research (BFR) (in Swedish).

Cold, B. (1989) Om arkitektur og kvalitet, *Tidskrift för Arkitekturforskning*, pp. 31–46 (in Norwegian).

Fall, Å. (1999) *Steg för steg*, Gothenburg: Chalmers University of Technology (in Swedish).
Hurtig, E. (1995) *Hemhörighet och stadsförnyelse*, Doctoral thesis, Gothenburg: Chalmers University of Technology (in Swedish).
Jensfelt, C. (1978) *Ombyggnad enligt sammanjämkningsprincipen*, Stockholm: Swedish Council for Building Research (BFR) (in Swedish).
Jensfelt, C. (1994) *Utformning av byggnader ur förvaltningssynpunkt – förslag till utvecklingsprogram*, Stockholm: Royal Institute of Technology (in Swedish).
Jensfelt, C. (1996) *Byggnadsutformning, nyttjande och förvaltning, fallstudie i Skogås*, Doctoral thesis, Stockholm: Royal Institute of Technology (in Swedish).
Junestrand S. (1998) *IT och bostaden ett arkitektoniskt perspektiv*, Doctoral thesis, Stockholm: Royal Institute of Technology (in Swedish).
Lang, A. (2001) *Personal communication with architect Arne Lang*, Gothenburg Community.
Lawson, B. (1991) *How Designers Think*, London: Butterworth Architecture.
Lundgren, O. (2001) *Personal communication with manager of the business department*, Bostad AB Poseidon.
Merriam, S. (1994) *Fallstudien som forskningsmetod*, Doctoral thesis, Lund: Lund Institute of Technology (in Swedish).
Niklasson, J. (2001) *Personal communication with project leader Johan Niklasson*, Bostad AB Poseidon.
Rönn, M. (1989) *Att projektera med hänsyn till underhåll och arbete*, Doctoral thesis, Gothenburg: Chalmers University of Technology (in Swedish).
Rybczynski, W. (1988) *Hemmet*, Stockholm: Bonnier (in Swedish).
Solberg, G.B., Nylander, O., & Forsman, P. (2001) *Tidningen Boendekvalitet nr 6*, Stockholm: Tenants' National Association, pp. 3–4 (in Swedish).
Starrin, B. & Svensson, P.G. (eds) (1996) *Kvalitativa studier i teori och praktik*, Doctoral thesis, Lund: Lund Institute of Technology (in Swedish).
Statistics Sweden (2001) *Bostäder och byggande*, http://www.scb.se [21 December 2001].
Sweeney, D. (1996) *Facilities Management*, London: E. & F.N. Spon.
Vidén, S. (1990) *Bättre bostadsförnyelse-sammanställning och slutsatser av 19 R&D project*, Karlskrona: Swedish Board of Building, Planning and Housing (in Swedish).
Werne, F. (1987) *Den osynliga arkitekturen*, Gothenburg: Vinga Press (in Swedish).
Werne, F. (1993) *Böndernas Bygge*, Höganäs: Wiken (in Swedish).
Williams, B. (1996) *Facilities Management*, London: E. & F.N. Spon.

Chapter 23
Conclusions

Brian Atkin, Jan Borgbrant & Per-Erik Josephson

Ways forward to construction process improvement

The construction process has been examined in all three of its product-related phases: definition, manufacture and use. Each of 21 main chapters, and the projects they report, address a specific area of need. They are supported, in many cases, by extensive literature reviews to establish the state-of-the-art and from that draw out evidence to support the particular research problem that is being addressed.

Awareness of literature published at home and internationally has been a necessary part of this activity: rarely, if at all, are problems unique to a country or region. Even if they were, valuable insights into problem solving would be lost if surveys looked no further than the national boundary. The contributors to this work have shown they are accustomed to searching for, and finding, evidence from diverse sources to support their case. Furthermore, they have provided the reader with access to literature from another culture and language that would otherwise be inaccessible to most. For that, they are to be commended.

Swedish construction has a concern about its future – a sentiment that is echoed in many other countries – and is determined to concentrate its resources into making strategically significant breakthroughs. One of the goals of the *Competitive Building* programme is to raise the competence of the sector's workforce; another is to create agents of change – people who are able to think strategically, while applying themselves to practical ways of improving the efficiency of the construction process. Although there is much still to be done, the results so far point to some measure of success towards these goals. In essence, collaborative research between university-based teams and active units within companies has been reported – in some cases, in much detail – and shows how co-operation works in practice.

Identifying the most relevant and potentially fruitful areas for *breakthrough* research is something that has to be done together. Even so, a crucial question is how the results of the research will be transferred and implemented in the companies and the sector as a whole. *How well does the research described and discussed in the chapters really address current and future problems?*

Characteristics of the research projects

Research projects can be classified in a variety of ways, but only occasionally, if at all in some fields, are they assessed collectively against criteria that reveal their contribution (evidential or potential) to science and industry. A national programme, for which there is focus and influence over resources and outputs, has the opportunity to undertake this kind of assessment; indeed, it must.

For our purpose in this assessment, we have classified the research projects reported in this work in terms of their orientation towards technical or human aspects, product or process, academic or practical aspects and strategic or operational aspects. The approach adopted in any given project depends on the problem defined, indeed how it is defined, the researcher's background and experience, and the kind of outcome or specific results that the sector is seeking. We now examine how the projects are oriented and discuss what this means for improving the construction process and, thus, the sector's prospects for greater competitiveness.

Technical versus human oriented research

Technical research has a focus on questions about materials, technology and construction. Examples would include improving the performance of concrete and extending existing structures. Human oriented research has a focus on questions in which knowledge and values, among other things, are needed to be able to work with materials and techniques. Examples would include dealing with relationships between the actors within a novel procurement method and individual behaviour. In some projects, there is a focus on both, but in general just one aspect is in focus.

Product versus process oriented research

Product research has a focus on the product, be it an apartment, house, factory, floor or comfort conditions. Process oriented research is focused on the methods, procedures or courses of action needed to realise a product. The emphasis is upon *what* shall be done and *how* it shall be done. Examples include *how* the actors communicate, *what* type of tools they use to make the communication effective, and *how* does experience feed back and learning take place? Most

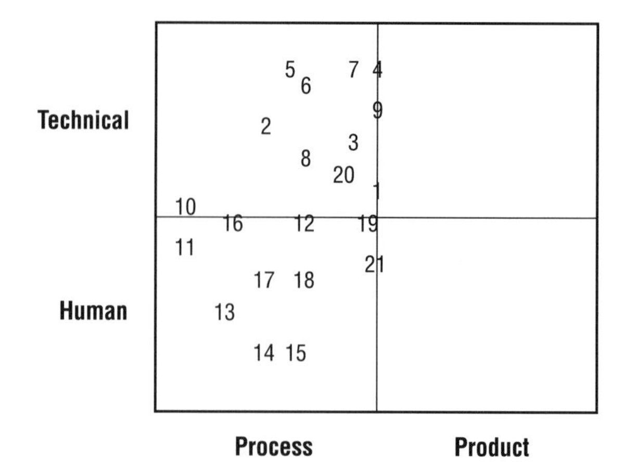

Fig. 23.1 Research projects classified according to their problem focus.

of the projects have a focus on process, but some of them forge a close link between the process and the product. (See Fig. 23.1.)

Academic versus practice oriented research

Academic research is focused on discovering new knowledge. The researcher can be searching for new formulations or connections between materials or other substances. The end results are often found in other research projects, where the new substance is critically analysed in order to find new applications. Practice oriented research focuses on questions that must be solved for the sector, in particular, producers of goods or services. It may be to do with finding a better tool or designing a new service. Evaluation of the results takes place in a practical setting. Most of the projects are practice oriented.

Strategic versus operationally oriented research

Strategic research is focused on questions that are linked to general political aspects, management issues and topics of interest to various disciplines. Examples would include *how* to position a company in the marketplace to take advantage of new or untapped demand and streamlining information management systems. Operationally oriented research has more to do with questions that are linked to work done on the construction site or in the factory. Examples would include improving materials handling and how to take accurate measurements for assessing performance of some kind. Within the projects is a diversity of approaches, some touching on both strategic questions

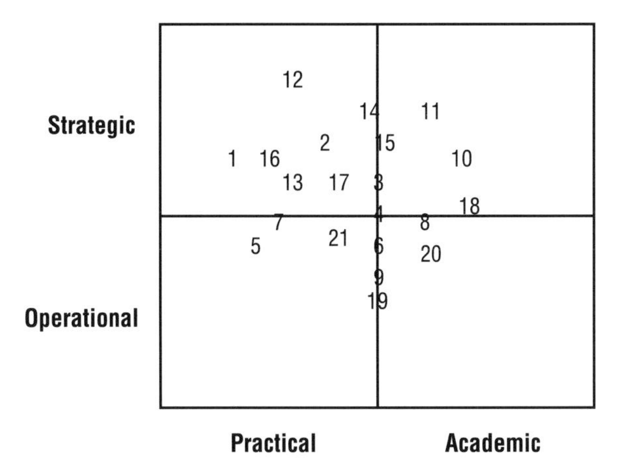

Fig. 23.2 Research projects classified according to their results focus.

(i.e. *what* should be implemented) and operational questions (i.e. *how* it should be implemented).

Looking across the 21 projects, we can conclude that the focus for each project is different, with different aspects emphasised according to the problem definition. This implies that, generally, the research results have the chance to be implemented in strategic, as well as practical, operational work, and that in doing so improve efficiency and the prospects for greater competitiveness. Moreover, an academic or scientific approach can be used to solve a practical problem on a construction site: not only that, it may be the only way that can lead to a workable solution. (See Fig. 23.2.)

Linking scientific research with the goal of industrial competitiveness

The approach taken in guiding the research, and in reporting on it here, has involved setting five tests or criteria. These were discussed in various respects in the Introduction and are summarised below, in terms of satisfaction with:

(1) the values for which we (the editors) stand, i.e. *what* we regard as rigorous and testable;
(2) the thoroughness with which the subject has been surveyed, i.e. *where* the researchers have looked;
(3) clarity in defining and delimiting the problem, i.e. *how* well drawn the problem is that is driving the research;
(4) transparency of the design of the research, i.e. *why* the research is being performed in a particular way; and
(5) critical analysis of the results, i.e. *when* and *how* the companies can use the results.

In reviewing the above, in the context of the research reported here, we can see evidence of different scientific and theoretical foundations or approaches; for example, natural science, basic and applied research, positivism, hermeneutic, social science and behavioural science. Each of these must be appropriate to the problem being solved. Also, each requires different criteria for assessing scientific worth and suitability.

Generally, we can say that an applied research approach often succeeds at the implementation stage, or rather it is easily facilitated, as it raises the prospect of practical benefits (for the companies) and, with it, brings improvement to the construction process overall. This does not mean that other approaches are unlikely to achieve the same, but they may require further research to extract results that the sector can use directly.

One reason why satisfactory results are achievable is that the research has been carried out from a distinct, scientific perspective in a transparent and controllable way. Not all of this can be derived from the basic skills that the researchers possess. Other steps must be taken. For the *Competitive Building* programme, the researchers' achievements cover, not only the results of their research, but also the acquisition of knowledge, skills and tools, much of which is derived from progressive, formal studies. The overriding consideration has been, therefore, that the researchers should, in every respect, strive for the highest scientific (academic) level. To do this, they must be equipped for the task in hand. Considerable effort has gone into providing the researchers with the key to unlock the door that leads to a deep understanding of the problem in front of them and that also prepares them for others to come. This understanding is vital.

Implementation of the results in order to improve competitiveness

The type of application, or area of application, that we might face depends on how well the problem definition is formulated. In the case of especially practical applications, the main concern in the formulation is how to facilitate implementation of the research results. In the case of problems that have a more academic basis, there will be work to do in transforming the research results into a form where they can be successfully introduced into the practical world. Naturally, the inherent character of the group that will share in the research results has a decisive influence on how the results are handled, or should be handled, by the company concerned.

Problem definitions that are well supported by essential theory can lighten the task of putting the results into practice, where the recipients are people who routinely work with matters of strategic importance. However, in the case of people who work at the operational level, there may have to be some adaptation and further work to complete the information needed to enable the results to be implemented. This may involve taking examples from daily practice and using them to show *where* and *how* the research results fit in. This means providing robust descriptions of the changes that the results necessitate in terms of different ways of working and forms of communication.

Research results that require changes in power structures can be taken to mean quite the opposite of what was intended, depending upon the different perceptions of the different managers, professions and trades involved. Research results that the company's senior management understands and is prepared to act upon may not be of much use or interest to middle management, where such results may be greeted with considerable scepticism. Furthermore, results that are worth much to specialists, for example, designers and those in other technical disciplines, can be regarded as entirely unrealistic by skilled workers on the construction site.

Answers that help deal with questions on the implementation of research results within a company are the responsibility of line managers and the researcher in question. The researcher's responsibility is, in particular, to communicate the results and the principal consequences these imply for the company. The company's management, including line managers, is responsible for determining the impact the research results have on their activities and must attend to the changes that have to be made. Additionally, the researchers must, within their project, incorporate an effective plan for implementation that can bring out real gains for the company: this must not be regarded as an afterthought. In the case of research supported by public funds, there will be a duty to disseminate the findings widely. The company must be alive to this and accept it as a formal commitment in return for the public's funds and act of good faith. It is of natural concern, therefore, that research results are available in reports and papers, as well as embodied in the researchers as bearers of new thinking and know-how. Companies must recognise this new-found knowledge and the skills that the researchers can bring, and be able to use them positively and creatively.

Having taken the metaphorical key and used it to unlock the equally metaphoric academic door, the researcher must use another key to unlock the door that leads into the company's world.

A holistic view of research design

Designing research projects is a subject in its own right and for which there is little space here. Even so, it is important, once again, to stress the importance of problem definition. Improving the efficiency of the construction process and, thereby, raising competitiveness within the sector, requires that attention be focused on problem definition. For companies, this will ensure that their long-term aspirations materialise through the results of relevant research.

From the research covered in this work by the individual researcher contributions, it is possible to see that the *Competitive Building* programme has achieved some results of its own, notably closer ties between the universities and industry. Implicit in our approach, as alluded to in the Introduction, has been to ensure that problem definitions are formulated precisely and in co-operation between the researchers, their academic supervisors and representatives of the companies (their industrial mentors). The design of the research projects and the work plans fleshed out from them are aimed at strengthening those ties. The applied aspects of the research, in every sense of the term, have been significant.

Implementation, that has been a major theme of this concluding chapter in reflecting the aims of the *Competitive Building* programme, is the ultimate key for everyone concerned. The test of good research of the kind promoted in the programme is unquestionably about delivering results that the sector can use and for which there is a clear and traceable path leading back to sound scientific principles, theories and methods.

The achievement of these goals is explicit in the research plans for the individual projects. Good research design means that problems are correctly defined and the most appropriate method or methods for finding solutions or, at least, answers that lead ultimately to solutions, are selected. But in the end, the real key to improving the construction process and, thus, competitiveness in the sector is collaboration between researchers and practitioners; that is, in this case, between universities and companies. The link between the emerging research results and the matter of competitiveness has not been proven, but neither has it been refuted. Time will tell whether the research results lead to an improved process that, in turn, leads to greater competitiveness.

Index

2D CAD 256
30-year reconstruction 283
3D CAD 256, 266
4D CAD 260, 266–76
  benefits 270–1
  limitations 271–2

academic oriented research 294
action research 238
after sales 23
agents of change 2
agents-mechanism-effect 76
agile production 36
air flows 85
alliance 145
American Productivity and Quality Center, The (APQC) 220
ArchiCAD 259
architects – writings of 240–43
architectural
  attributes 278–90
  design 240–53, 256
  innovation 159
  quality 2
  theory 242
auditing process 230
audits 217, 230
authority 214
AutoDesk 258
automobile industry 19

BAS • CAAD 246–7
benchmarking 190, 199
best practice 225
BOOT 135, 156
BOT 135, 156
BPR 213, 215, 226
BSAB 259
BTO 156
building
  and operations management 5
  codes 107
  failure 82
  material 4
  material market 183
  physics 68, 82–3, 106, 112
  related illness 94
  research 1
business
  excellence models 236
  goals 213
  performance 226
  processes 213
Byggkostnadsdelegationen 31

CAAD 240
  development 241
  ideas 256–8
  research 241
  systems 241–2
  tools 255
CAAD model 258, 262–3
  building 258–60
  exporting documents 263
CAAD modelling
  system 258
  tool 260
CAAD programs 241, 242, 255
  prospects 248–53
  shortcomings 245–8
CAD 38, 255
  program 246
  research 240
  techniques 256
capillary protection 78
capital 4
  goods 146
case study 162, 231, 274–5
cast in situ concrete 118, 119–20
change 288
CIFE 272
client
  awareness 108
  needs 108
  objectives 205–206
  requirements 32, 108, 109

CO_2 emissions 31, 39
coach 214
commercial
  building 44, 195–209
  environment 4, 5
  leadership 213
commitment 109, 145, 214, 227
  of people 215, 216
communicating project concepts 255–65
communication 4, 5, 108, 144, 145, 151, 187, 205
  channels, 145, 220
  technology 69
  web 197
comparative study 161
competence 144, 185, 187
competition in tendering 131
competitive advantage 208
competitiveness 130, 139, 143
  improve 296–7
component standardisation 20
computer
  industry 19
  modelling 252
concrete 118
CONNET 70
construction
  contracts 131
  innovation 146
  process improvement 292–8
Construction Industry Council, The (CIC) 157
Construction Primer, The 70
consumer uncertainty 198
contextual uncertainty 199, 207
continual improvement 4, 5, 214, 215, 217, 225
  culture 237
continuous learning 145
contractual
  problems 145
  relations 130, 226
control units 146
cooling reduction of air 78
co-operation 143, 144
  cross-functional 187
co-ordinated standard approach 133, 136
co-ordination of design and building processes 106
CoPS 146
core values 226–7
cost
  reduction 91, 183
  savings 155
creating decision support 255–65
cross-functional
  co-operation 187
  groups 191
culture 185
customer
  demands 4
  expectations 215
  focus 211, 214, 215, 227
  needs 213, 214, 215
  orientation 2
  relations 215
  requirements 22, 214
  satisfaction 4, 200, 214, 215, 217, 236
  use 4
  value 214, 215
customer-focused design 3
customisation of manufactured housing 15

damage 93
DBFO 135, 156
DCMF 156
decision
  support 242
  support during design 68
  support tool 217
  tools 68
defects 93, 96, 213, 216, 274
Dependency Structure 24; *see also* Design Structure Matrix
design 23
  activity 242
  and construct approach 133–4, 136
  criteria 106, 109
  goals 242
  methodology 241
  object classes 249
  practice 106
  research 240–53
  team 38
  tool development 240–53
Design Precedence Matrix, The 24; *see also* Design Structure Matrix
Design Structure Matrix, The (DSM) 24
development environment 188
Development Fund of the Swedish Construction Industry, The (SBUF) xiii
deviations 216
diagnosing damage 98
disciplined integrated problem solving 197
divided contract approach 132, 136
durability 68, 287–8
Dutch Foundation for Architects Research, The (SAR) 36
dynamic
  capability 228
  team 187

embodied energy 44
employees commitment 185
end user 108, 146
energy
  conservation 48–50
  efficiency 68
  performance of buildings 44
  transfer 106
  use 106
ENEU 59
ENORM 45
environmental
  assessment 101
  awareness 211
EPD, Environmental Product Declaration 35, 41

EQUAL 137
errors 213
ethnography 161
European Parliament, The 44
Europeanisation 15, 28
evaluation of tenders 137–9
execution 289
existing buildings 93
experience 185
experiment 161
explicit knowledge 184
external
  audit 235
  collaboration 145
  environment 195

facilities management 5, 38, 278–90
fact-based decisions 227
factual decision-making 215, 217
feedback
  loops 146
  of information 38
FEM Design Plate 125
Field and Laboratory Emission Cell (FLEC) 103
financial
  commitment 145
  risk 157
  service industry 19
financially free standing project 135
Finite element methods (FEM) 125
future emissions 93, 96–7
FutureHome 15

Greenzone 46–54
Grounded Theory Method 286
group processes 176

Hand Arm Vibration Syndrome (HAVS) 120
health
  aspects 61
  check 235
healthy indoor climate 2, 93
heat flows 83–5
HEAT2 85
HEAT3 85
Hett 125
high performance concrete 118, 121–2
high-energy consumptions 106, 112
High-Performance Commercial Buildings 84
high-rise house building 119
histories 161
HK-BEAM 70
Hong Kong Tunnel 156
housing 31
human oriented research 293

IDEAS 267
IDEF0 method 221
IFC 258
implementation of research results 296–7
improvement processes 225–37
incremental innovation 159, 227
individual performance 211
indoor
  air problems 93
  air quality 58, 71, 94
  climate systems 56, 60–61
industrial competitiveness 295–6
industrialisation 2
industrialised
  building 3
  production method 4
infiltration water protection 78
information 145
  management 270–72
  quality 211
  success model 220
  system success 219
  technology 69, 220
initial cost 51
innovation 143, 154, 157–60, 228
  key factors 148
  principles 159
  role of 157
  theory 143, 144
  types 159
innovative capabilities 144
innovativeness 143
integral building envelope performance assessment 72
integrated design 33
  and production 4
integrated life cycle design 31, 37
  methods 32, 38
inter-firm alliances 145
internal audit 229, 235
internal auditors 230
internal environment 214
international competition 183, 211
International Energy Agency, The (IEA) 33
International Performance Simulation Association, The (IBPSA) 71
internationalisation 211
Internet 219
inter-organisational collaboration 176
inter-organisational communication 145
inter-organisational processes 219
intra-organisational communication 145
involvement of people 215, 216
I-SEEC 70
ISO 14000 35, 235
ISO 7730 63
ISO 9000 213, 215, 225, 235
ISO certificate 235
IT solutions 38

Japanese manufacturing philosophies 106
JIT 226
joint ventures 135

Kaizen 215
know-how 4, 143
knowledge 144, 184–5, 186, 213
  bearers 186

creation 185
management 184, 186, 215
transfer 186

lack of marketing skills 145
LCA, life cycle assessment 32, 60
LCC 32, 34, 44, 45, 83
analyses 50, 59, 60–61
approach 56
calculations 41, 62–3, 64–5
principles 58–9
research 58–9
techniques 34
tools 58–9
use 58–9
leadership 187, 198, 214, 215, 235
lean enterprises 215
lean production 36, 213, 215
learning 4, 5, 144, 145, 151, 183–93, 230
arena 187, 192
capacity 187
double-loop 184
environment 192–3
hinder 147
organisation 184, 185, 213, 215
processes 185, 228
single-loop 184
life cycle design – tools 38
life cycle inventory 35
light-gauge steel framing 269–70
benefits 269–70
logistics planning 268
loss of project control 145
loss of technology 145

M4I, Movement for Innovation 31
maintenance cost 52–3
management
behaviour 226
commitment 227
information systems 211–23
information systems success 220
managing housing 279–82
manufacturability 20
market 143
control 2
demands 15
intelligence 198
marketing proficiency 203–205
mass-customisation 36, 37
mass-production 37
material 289
properties in situ 101
meetings 191
MISTRA Research School of Sustainable Building 101
modular
architecture 20
design 18
innovation 159
products – benefits 21
system 19
Modular Functional Deployment (MFD) 22
modularisation 15
methods 22
modularity 19
module drivers 22
profile 23–4
MOIST 85, 88
moisture 106
control 268
damage 99
design 100
diagnosis 99
flows 86–8
measurement 71
mechanical behaviour 100–101
penetration 87
problem 112
status analysis 95
monitoring 229
motivation 109
multiple criteria decision air procedure 72
Museum of Modern Art, Stockholm, The 79
mutually beneficial supplier relationships 215, 219

National Institute of Public Health, The 69
NCC Housing's Total Package Concept 221
networks 146, 176, 225
new concrete material technology 120–21
new products 144
NHER Evaluator 85
NICC 258
North Sea oil projects 155

object-oriented modelling 252
observation 231
open building 17, 18, 26, 28, 32, 36–7, 41
open industrialisation 26
operational energy use 44
operational uncertainty 199, 207
operationally oriented research 294
optimal building performance 106
organisation 4, 5
change 144
communication 176
cost 52
development 212, 213
dilemma 198
learning 186, 230
process-orientation 211
stability 225
success 183
Our Way of Working 234–5

participating researcher 231
partnering
critical views 173–4
external environment 174–5
role of contracts 174–5
selection process 174–5
structures 168–180
partnership solutions 154, 253

Partnerships Victoria Guidance, The 157
patent 199
paying client 108
performance
  contracting 134, 136
  indicators 68
  measurement 229
  measurement system 234
  requirements 107
personnel 4
PFI (Private Finance Initiative) 157
physical status 98
P-label 70
planned design changes 23
planning 290
PPP (public-private partnerships) 135, 154–65
practice oriented research 294
pre-assembly 17
prefabricated concrete structures 39
primary emissions 97–8
private sector 154
Problem Solving Matrix, The 24; *see also* Design Structure Matrix
problem-solving concept 198
process
  approach 215, 216
  improvement 3
  innovation 158, 215
  map 216
  orientation 211, 212–19, 220
  oriented research 293
  quality 211
  transparency 211–23
process-oriented organisation 211
procurement 130, 143
  method 195
  systems 136
product
  architecture 20
  change 20
  characteristics 33
  definition 4
  development management 20
  development processes 195
  diversity 225
  families 18, 19
  manufacture 4
  modularisation – methods 27
  oriented research 293
  performance 20, 199
  quality 219
  requirement 217
  use 4
  variety 20
product platforms 18, 20
  concept 19–20
  creation 21
product value
  assessing 199–200
  chain 212
production 23
productive development 32, 197–8
productivity 2, 183
project
  concepts 242
  efficiency 211–23
  management 107, 192, 228
  objectives 108
  outcomes 200
  performance 211
  success 113, 115, 154, 212, 236
  team 183–93
public sector 154
purchasing 23
  behaviour 183

QFD, Quality Function Deployment 32, 33, 108
quality 23
  awareness 226
Quality Council of the Building Industry, The 69
quantity surveying 38

radical innovation 159
rational plan 197
rationalised real estate redevelopment 5
real estate 4, 195
recycling 23
reduce cost 183
refurbishment 195–209
relationships 144
renewal 288–9
research
  academic oriented 294
  characteristics 293
  criteria 6
  design 297–8
  human oriented 293
  operationally oriented 294
  practice oriented 294
  process oriented 293
  product oriented 293
  strategic oriented 294
  technical oriented 293
researcher education 2
residual service life 99–100
resource uncertainty 198
reward 144, 145
  system 228
risk 82, 95, 145, 154
  capital 144, 145
  distribution 157
  transfer 157
RISK1 87
room air distribution 58
root cause 214

Swedish Council for Environment, Agricultural Sciences and Spatial Planning, The (FORMAS) xiii
schema evolution 249
science of finance 19
secondary emissions 98
self-assessment 217, 236
self-compacting concrete 118, 122–3

self-generating process 228
services 146
services sold to the private sector 135
SETAC 35
sick building syndrome 79, 94, 106, 112
skill 187
social change 144
social pattern 211
socialisation 186
software package 146
Swedish Foundation for Strategic Research, The (SSF) xiii
standard approach 132–3, 136
standardisation 17, 18
STEP 258
strategic oriented research 294
strategic planning 216
structural frames 123
styling 23
suppliers 146
supply 4
  chain 212
  chain management 36, 213, 215
sustainable development 2
Swedish Building Cost Commission, The 123
Swedish environmental law 79
Swedish National Testing and Research Institute, The (SP) 69
symbols 249
SYRE 259
system
  approach to management 215, 216–17
  design skills 146
  innovation 159
  quality 211
  theory 150
  use 211

tacit knowledge 183
target-oriented teams 187
task uncertainty 198
team
  composition 193
  learning 186–8, 192
  members 187
teamwork 187, 215, 226
technical
  criteria 108
  innovation 158
  oriented research 293
  proficiency 202–3
  specification 23
technological
  competence 143
  errors 115
  evolution 23
  uncertainty 198
temporary alliance 225
temporary coalitions 146
tender – competition 131, 139
territorial thinking 187
TorkaS 125
TQM 213, 215, 225, 226
  adoption 225–6
Treasury Taskforce, The 156

UK National Audit Office, The 156
uncertainty 145, 195, 198–9
  levels of 198
unhealthy buildings 94
US Department of Energy, The (DOE) 84
user
  requirements 146
  satisfaction 211

value 143
  chain 211, 219
  chain management 212, 215
values 185
vapour protection 78
variance 23
ventilation
  controls 58
  principles 57–8
  systems 58
vertical extensions 266–76
vibration moment 124
volatile organic compounds (VOC) 94

weather-tightness 268
white fingers 120
whole life costing 24–5, 34
whole life cycle cost 53–4
wicked problem – characteristics 243–4
work
  performance 61
  procedure 235
working space 268
writings of architects 240–43
WUFI 87